Le

on Brachet

Aix-les-Bains in Savoy

The medical treatment and general indications

Le

on Brachet

Aix-les-Bains in Savoy
The medical treatment and general indications

ISBN/EAN: 9783337314514

Printed in Europe, USA, Canada, Australia, Japan

Cover: Foto ©berggeist007 / pixelio.de

More available books at **www.hansebooks.com**

(*IN SAVOY*),

THE MEDICAL TREATMENT
AND
GENERAL INDICATIONS.

BY

DOCTOR BRACHET,

Physician to the Bathing Establishment of Aix-les-Bains and Marlioz; Senior Physician of the Thermal Hospital; Honorary Physician of the Anglo-French Asylum; Physician to the P. L. M. Railway Company; to the Society of Artists and Inventors of Paris; Member of the Hydrological Society of Paris; Member of the Medical Academy of Rome; Member of the Academy of Savoy; and Member of the Medical Societies of Geneva, Montpellier, &c. &c.; Chevalier of the Legion d'Honneur; Commander of the Royal Order of the Saviour (Greece); Commander of the Royal Order of Isabella (Spain); Officer of Nichau; Chevalier of Charles III. (Spain).

SECOND EDITION.

HENRY RENSHAW,
356 STRAND, LONDON.
1891.

[*All rights reserved.*]

Ballantyne Press
BALLANTYNE, HANSON AND CO.
LONDON AND EDINBURGH

TO THE

Distinguished Members of my Profession

IN

GREAT BRITAIN AND IRELAND,

FROM WHOM

I HAVE RECEIVED VALUABLE INFORMATION, WITH MANY
PROOFS OF CONSIDERATION AND KINDLY FEELING,

THIS LITTLE WORK

IS INSCRIBED AS A SINCERE THOUGH INADEQUATE MARK
OF RESPECT AND ESTEEM BY

THE AUTHOR.

PREFACE.

Since the publication of the first edition of my clinical study of the mineral waters of Aix-les-Bains and Marlioz, the increased importance of these spas, and my own wider experience, enable me to furnish more copious notes.

Situated on the borders of Lake Bourget, at the foot of the Alps between France, Italy, and Switzerland, Aix is easy of access by the railway from Paris to Rome.

The English visitors, appreciating natural beauties, have almost transformed it into an English colony, especially during the spring and autumn months. Her Majesty the Queen Victoria and several members of the English Royal Family, the Emperor of Brazil and his Majesty the King of Greece, have paid repeated visits to Aix-les-Bains of late years, and have naturally contributed to the increased influx of visitors. Last year (1890) the bathers numbered thirty thousand.

I do not pretend to enter into details of physiological or clinical research, but simply to indicate certain practical results of my daily observation during the last twenty-five years, which I trust may be favourably received by my colleagues who may not be acquainted with the valuable properties of the waters of Aix-les-Bains and Marlioz. With a view to facilitating the excursions of patients, and enabling them to see the most of this beautiful country, I have given details for their guidance which would not otherwise have found a place in a medical work.

L. BRACHET.

AIX-LES-BAINS, SAVOIE, FRANCE,
  March 1, 1891.

CONTENTS.

	PAGE
PREFACE	vii
HISTORICAL RETROSPECT	1
CLIMATE	4
GEOLOGY	5
THERMAL ESTABLISHMENT	6
CHEMICAL ANALYSIS OF THE WATERS OF AIX-LES-BAINS	9
MODE OF USING THE WATERS AT AIX	11

Internal Use—Douche—Massage—Steam Baths (bouillons—Berthollet)—Baths—Swimming Baths (Piscines)—Inhalations—Spray.

GENERAL AND THERAPEUTIC ACTION OF THE WATERS	16

Thermal Fever (*Poussée*).

DURATION OF THE TREATMENT	18
BEST SEASON FOR THE COURSE	18
DIET DURING THE COURSE	19

DISEASES SUCCESSFULLY TREATED AT AIX-LES-BAINS.

RHEUMATIC DISEASES	20
RHEUMATOID ARTHRITIS	25

Rheumatic Gout—Chronic Rheumatic Arthritis—Nodosity of the Joints—Nodosity of Heberden—Arthritis Deformans.

CONTENTS.

	PAGE
GOUT	30
NEURALGIA—SCIATICA—CASE OF GOUTY SCIATICA	42
DISEASES OF THE ORGANS OF RESPIRATION	45

    Rhinitis—Ozæna—Pharyngitis—Catarrh of the Pharynx—Sore Throat—Granular Pharyngitis—Clergyman's Sore Throat—Herpes of the Pharynx—Rheumatic and Gouty Sore Throat—Chronic Laryngitis—Chronic Bronchitis—Hay Fever—Asthma—Catarrhus Senilis—Winter Cough—Phthisis.

SYPHILITIC DISEASES	54
SCROFULA	57
SKIN DISEASES	59

    Psoriasis—Prurigo—Eczema—Herpes—Acne.

WOMEN'S DISEASES	62

    Simple and Follicular Vaginitis—Hæmatocele—Uterine Diseases: Amenorrhœa—Dysmenorrhœa—Metritis—Ulceration—Leucorrhœa—Uterine Tumour—Disturbances at the Change of Life.

HYSTERIA	70
ANÆMIA AND CHLOROSIS: CASES	71
CRETINOID MYXŒDEMA (CACHEXIE PACHYDERMIQUE OF CHARCOT)	73
PARALYSIS	78

    Local Paralysis—Hysterical Paralysis—Mercurial and Lead Palsy—Syphilitic Paralysis—Hemiplegia—Paraplegia.

SURGICAL CASES	83
ILLNESSES FOR WHICH THE WATERS OF AIX ARE UNADVISABLE	85

    Torpid Phthisis—Cancerous Diseases—Some Cases of Heart Disease—Nervous Irritability—Congestion or Softening of the Brain.

ADJUNCTS TO THE TREATMENT OF AIX.

	PAGE
SPRINGS OF MARLIOZ	87
CHALLES WATERS	89
ST. SIMON WATERS	91
WHEY CURE	91
ELECTRICITY COMBINED WITH THE TREATMENT OF AIX-LES-BAINS	92

PART II.
GENERAL INDICATIONS.

ROUTE BETWEEN PARIS AND AIX-LES-BAINS	97
GENERAL COUP-D'ŒIL OF THE COUNTRY	97
THE TOWN OF AIX-LES-BAINS	100
HOTELS—BOARDING-HOUSES—LODGING-HOUSES	101
CARRIAGES—OMNIBUSES—RIDING HORSES—DONKEYS	101
STEAMERS AND BOATS	102
TARIFFS	102
DIVINE SERVICE	104

Roman Catholic Church—English Protestant and Presbyterian Churches.

HOSPITALS	104

Thermal Hospital—Anglo-French Protestant Asylum.

ANTIQUITIES OF AIX	105

Arch of Campanus—Roman Baths—Temple of Diana—Sundial.

MUSEUM AND LIBRARY	106
AMUSEMENTS	107

Casino—Villa des Fleurs—Pigeon-shooting—Races.

	PAGE
WALKS	108

Roche du Roi—Boulevard des Cotes—Park of Marlioz—St. Simon—Pont Pierre—Château de Syllan—Maison du Diable—Bois Lamartine—The Grand Port—The Petit Port—Hill of Tresserves—Mouxy—Trevignin—Clarafond.

DRIVES 111

Waterfall and Tower of Gresy—Moulin de Prime—Grott de Bange—Tour de Lac—Castle of La Serraz—Castle of La Motte—Le Bourget et le Col du Chat—Castle of Bourdeau—Haute Combe—Castle of Chatillon—Drive to the Rhone by the Canal of Savière—Val de Fier—Mount Joigny—Valley of the Beauges—Castle of Miolan—Pierre Chatel—The Grande Chartreuse—Lake of Aiguebelette.

MOUNTAIN ASCENTS 118

Dent du Chat—The Revard—The Dent du Nivolet—Semnoz Alps—Tower of Cessens.

CHAMBÉRY 120

Les Charmettes—Challes.

ANNECY 122

AIX-LES-BAINS
(SAVOY).

HISTORICAL RETROSPECT.

The **strategical position** of Savoy, at the foot of the Alps, has from time immemorial given so great an importance to this country, that nothing could occur in Europe without its being concerned. Hence the interest attached to its history, which, apart from all patriotic enthusiasm, must be regarded as one of the most admirable and attractive.

The *savant* may pursue his pre-historic studies, from the summit of the mountain glaciers to the bottom of the lakes.

The historian discovers that in remote ages this small spot, so favoured by Nature, was successively occupied by the Allobrogi* and by the Romans, the masters of the world, whose bathing propensities are well known. They constructed at Aix an immense establishment, replete with luxury and comfort. Abundant proofs of the Roman occupation have been found at Aix, sufficient to reproduce the plan of the

---

* Many customs, still existing in Savoy, trace their origin to the occupation of the Allobrogi.

Roman baths, which closely resembled those of Titus, Agrippa, Antonine, and Diocletian, at Rome. I have found in the Baths of Caracalla, at Rome, and in the grounds at Aix, some pieces of brick and amphoræ completely identical. These baths were destroyed on the first invasion of the barbarians. Some historians assert that the Emperor Gratian restored them, for which reason Aix was called *aquæ Gratianæ*. I can discover no mention of the subject in the Latin writers.

Christianity had no share in the re-establishment of the Baths. For a period of six centuries Aix, in common with other bathing places, disappeared from the pages of history. At the end of the fourth century it fell under the yoke of the barbarian; it next passed under the dominion of the Merovingian and Carlovingian kings; then under that of the Dukes of Burgundy. Ultimately, after 1032, it became successively the property of the counts, dukes, and kings of Savoy, whence those heroes (improperly termed by Thiers the wolves of Savoy) extended their possessions little by little, and founded a chivalric dynasty which reigned without interruption for nine centuries, the last illustrious representative placing on his brow the iron crown and ascending to the Quirinal. The old castles and abbeys surrounding the lake demonstrate the high estimation in which Aix and its valley have been held since the thirteenth century;* but it

---

* We strongly recommend visitors to attempt the ascent of the mountains, a task now rendered easy, as the study of their geological characters, with those of our grottoes and valleys, is extremely

was only since the sixteenth century that attention was again directed to the Baths, and Aix was constituted a marquisate. Henry IV., King of France, came to bathe there: for many years the appellation of Royal was given to the Swimming Bath, over which now passes a street bearing the monarch's name.

In 1739 Aix was destroyed by fire. King Victor Amedée III. created a suitable bathing establishment in 1776. During the twenty years of the first French occupation the Baths of Aix were neglected. Many plans and projects were contemplated, but political disasters impeded their execution. The return of the House of Savoy, in 1815, inaugurated an era of progress for Aix-les-Bains, which continued under the direction of the Drs. Despine till 1860, when Savoy passed for the second time into the hands of France. Since that epoch Aix has greatly increased, not suddenly, but in a gradual manner, owing to the superiority of its spas and to the expenditure which the Government has unhesitatingly incurred.

The proverb says—Woe to a people with too much history! The Savoyard feels no regret on that score; in the midst of these successive changes he has remained faithful to the flag of Savoy; always pas-

interesting: also occasionally to fish for *palaffites* (antiquities), in the lake of Bourget. This small lake contains four stations, or "Villages Lacustres," built on piles, at a long distance from the shore; probably used originally as storehouses, protected alike from thieves and wild beasts. Beautiful specimens of bronzes and flints of different periods, found in the lake, may be seen in the Museums of Aix and Chambery.

sionately attached to his mountains and his lakes; but fighting with equal bravery—and respecting equally all national institutions—during the alternate occupations of Italy and France.

CLIMATE OF AIX.

The town of Aix, situated in a valley, and protected from winds by the surrounding mountains, enjoys a mild and equable climate. This **topographical position** also accounts for the elevated temperature occasionally prevalent. As the majority of cases treated here result from the pernicious influence of sudden atmospheric changes, it is evident that the climate of Aix is especially adapted to rheumatic and gouty patients.

The rich and splendid **vegetation**—far superior to that of the adjacent countries—proves the excellence of the atmospheric conditions.

Patients exhausted by years of insomnia, experience sudden relief from neuralgic sufferings, with the benefit of soothing sleep. Unlike many towns in possession of sulphurous springs, and subject in consequence to disagreeable exhalations, Aix is entirely free from such a nuisance, the municipality having wisely expended enormous sums in disposing of the outflow of the Baths and in a perfect system of general drainage.

The average temperature is 50 degrees.

GEOLOGY OF AIX AND ITS ENVIRONS.

THE quantity and variety of **mineral springs** in Savoy have at all times interested geologists. Within the narrow space of a few miles are found the alkaline waters of St. Simon, the sulphureous waters of Challes and Marlioz, those of Allevard, Brides, and Uriage—in a word, all those waters now occupying our attention, in addition to innumerable springs abounding on all sides.

M. Mausson and Dr. Lombard, from Geneva, recognise that the surrounding mountains form the prolongation of the Jura Chain, and that they are formed of the same strata: Calcareous, Neocomian, Oxfordian, Coralian, and Salitic, found near the great Jurassic Chain, one of the outposts of the Alps. The valley of Aix presents numerous gaps, and the rises and falls are very frequent. In the Bauges you see how the Calcareous strata have been constantly raised. Probably the mineral springs pierced the soil after these disturbances. Considering the temperature of the waters of Aix, they must spring from a depth of about a thousand metres. They do not rise direct from the centre of the earth, but follow the intervals of the Calcareous strata of the Bauges. Thus in remote ages the inhabitants of this valley sank wells in their mountains, hoping to reach the hot springs.

M. Mausson recognises points of resemblance between the extreme north and south of the Jura. In fact, in the north are found the hot springs of Baden, the sulphuric waters of Schinznach, and the saline waters of Bermandorf, Wildegg, Millingen, &c. &c.; and in the south the innumerable springs of Savoy.

THERMAL ESTABLISHMENT.

1. The **alum and sulphur springs**, though differing but slightly, do not penetrate the soil at the same spot. The alum spring rises a little above the town, which it reaches by means of a gallery two metres in breadth and two in height, extending over ninety-two metres; at the extremity is the vast Serpent Grotto (so named from some harmless serpents found there), presenting magnificent stalactite effects. Beyond the grotto is the reservoir of the spring; here the water appears to be covered by a layer of grey and unctuous barégine, soft to the touch; huge gas bubbles occasionally appear on the surface. Beneath this layer the water is clear and limpid. Strangers must not fail to visit this grotto in order to be convinced of the enormous bulk of water of high temperature at the service of the baths, which can be used profusely, unlike those of Barèges and other places, where the limited supply renders great thrift necessary.

2. The sulphur spring rises in the Establishment; on raising the stone which covers it, numerous small bubbles of sulphuretted gas are visible on the surface.

The Establishment is composed of a ground floor (*soubassement*) and a first and second floor. Patients who walk with difficulty reach the first floor by

Etablissement Thermal d'Aix-les-Bains

means of an easy incline without ascending the staircase. Above a large wide staircase is the *Buvette* (Trinkhalle), a splendid hall, light and well-proportioned, separating the baths and piscines of the ladies from those of the gentlemen.

Those who cannot ascend to the *Buvette* find in the town several public fountains with taps of the mineral water. The maximum pressure of the **douches** obtainable on the ground floor is $15\frac{1}{2}$ metres, but the doucheurs have at their disposal a manometer, which enables them, by means of the different apparatus, to regulate the pressure of the douche according to the medical prescriptions.

The Establishment comprises :—

1. Two immense swimming baths with cold douches.
2. Two old swimming baths.
3. Two commodious family swimming baths with douches.
4. Forty-one single baths.
5. Twenty-five large douches with two doucheurs or doucheuses.
6. Twenty douches with a single doucheur or doucheuse.
7. Two douches *en cercle*.
8. Three douches *à colonne*.
9. Six vapour baths (Berthollet).
10. Two inhaling rooms.
11. Three rooms for administering spray.

12. Five vaporaria.
13. Six bouillons (steam baths).
14. Four ascending douches.
15. Four foot baths.

The maximum number of patients undergoing the various forms of **treatment** varies from 2000 to 2300 daily.

Recent additions and improvements relieve patients from the necessity of inconveniently early hours, formerly entailed by the inadequate accommodation, for their baths and douches.

The prices of the Baths are very moderate; and even the most luxurious douches cost only two shillings. Appended is a table showing the prices of the various kinds of baths.

TARIFF OF THE BATH,

Fixed by the Minister of Commerce, on April 26, 1882.

	Francs
Douches du soubassement et Annexe-Sud	2 50
Bouillon seul	1 50
Douches de l'Annexe avec bain	3 0
Douches des Princes vieux, neufs; douche neuve	1 50
Douche à colonne	2 0
Douches moyennes: centre, enfer, verticale, vaporarium	1 0
Douches Berthollet {après 6 h. du matin	1 50
{avant 6 h. ,,	1 0
Douches en cercle, en lame, locales, pharyngiennes, humage	1 0
Inhalation	1 0
Douches ascendantes	0 50
Bains de pieds	0 50
Bains réfrigérés en baignoires avec {après 6 h. du matin	2 0
ou sans la douche pulvérisée {avant 6 h. ,,	1 50
Bains ordinaires {après 6 h. du matin	1 50
{avant 6 h. ,,	1 0
Piscines, grandes {hommes	1 50
et petites. {femmes	1 25

Piscines ovale et carrée des Albertins .	.	. 0 50
Piscine de famille.—L'heure . . .	.	. 10 0
Visite de Grottes { jours ordinaires	.	. 0 60
{ jours d'illuminations	.	. 1 0
Portage, simple . . .	.	. 0 6
„ double . . .	.	. 1 0

THE CHEMICAL ANALYSIS OF THE WATERS OF AIX-LES-BAINS.

These mineral waters have been classed amongst **sulphur-soda springs,** and are distinguished from those of the Pyrenees by the proportion of carbonic acid gas and calcium bases which they contain.

We have said that two sources supply the Establishment of Aix, the one of sulphur and the other of alum; both have a temperature varying between 114° and 117° Fahr. The alum source (double sulphate of alumina and potash) was so called in former times when sulphate of alumina was designated as alum.

The **alum spring,** called the *gracieuse* by Daquin, is most employed for internal use, as it is easy of digestion. I give this explanation for the guidance of the numerous patients who fear to take the alum spring on account of its supposed constipating effects. These two thermal springs yield the enormous quantity of four million litres in twenty-four hours; to reduce their temperature natural cold water is added.

We use at least seven millions of litres daily for the medical treatment. Occasionally, after a severe storm, a slight reduction of temperature takes place, caused by temporary filtration of rain-water through the earth, but never sufficiently to interrupt the course of baths, or diminish the medical efficacy of the treatment.

ANALYSIS OF THE MINERAL WATERS OF AIX.

By M. WILM.

	Sulphur.	Alum.	
Free sulphuretted hydrogen	3·37 to 4·13 mm.	3·74 mm.	
Sulphur under the form of hyposulphite	3·84 mm.	3·60 mm.	
Carbonic acid gas	47·15 cc.	44·58 cc.	
(by weight)	0·0932 gramme	0·0882	
Nitrogen	13·03 cc.	12·5 cc.	
			0·1982
Calcic carbonate	0·1894	0·1623	
Magnesic ,,	0·0105	0·0196	
Ferric ,,	0·0010	0·0008	
Silica	—	0·0175	
Total deposit from boiling	0·2009		0·1983
Silica	0·0479	0·0365	
Calcic sulphate	0·0928	0·0810	
Magnesic ,,	0·0735	0·0493	
Sodic ,,	0·0327	0·0545	
Aluminic ,,	0·0081	0·0003	
Sodic chloride	0·0300	0·0274	
Calcic phosphate	0·0076	traces.	
Total constituents remaining in solution	0·2916		0·2461
Total fixed constituents determined	0·4925		0·4433

Sulphur Spring.   *Alum Spring.*
Organic matter very changeable.

Lithium	⎫			traces	⎫	
Potassium	⎬ 0·0050			doubtful	⎬ 0·0095	
Strontium	⎪			doubtful	⎪	
Iodine	⎭			traces	⎭	

Organic Matter of the Water of Aix, or Barégine.

The barégine of Aix, dried at 100°, leaves 50 per cent. of ashes, consisting, in 100 parts, of—

Silica	. .	37·41
Alumina	.	4·86
Ferric oxide	.	10·00 (about)
Substances undetermined { Hydrochloric acid, Sulphuric ,, Carbonic ,, }	.	11·76
Magnesia		small quantity—none
		100

MODE OF USING THE WATERS AT AIX-LES-BAINS.

Internal Use.—These waters are easy of digestion, producing neither sickness nor irritation, but only at first a feeling of aversion, which soon passes away. They promote and increase the urinary secretion and eliminate urea and uric acid. I prescribe them when there is no evidence of gastric or bilious symptoms. Occasionally, the treatment diminishes the action of the bowels, but it must not be suspended on that account; a gentle aperient, the tea of Aix, or a mineral purgative water, only becoming necessary. The exact quantity of sulphur water to be taken cannot be fixed, as it varies according to the individuality of the patient. Excess of the internal use of mineral waters, prevalent in many spas, is no longer tolerated; taken too freely they over-stimulate perspiration and the urinary functions; their digestion is impeded and absorption is thus neutralised. Certain diseases requiring a large absorption of sulphur may be advantageously treated by internal use of the Marlioz and Challes waters, both containing a great quantity of iodine, bromide, and sulphide of sodium. Three to four tumblers of Aix water daily generally suffice; but two glasses of Challes water are rarely exceeded.

Douche—Massage—Maillot—Vapour Bath —Piscines.—Every kind of douche can be obtained at Aix, local or general, ascending or descending, hot or reduced to any temperature, weak or strong, in light and spacious rooms. The patient sits down on a wooden chair, with his feet in hot water, and one or two doucheurs or doucheuses propel jets of water all over the body, hottest of all on the feet and legs.

Simultaneously, for several minutes, the doucheurs shampoo, rub, and knead every part of the body, thus stimulating the capillary and general circulation: the temperature, strength, and duration of the douches and shampooing are previously indicated by the doctor.

The massage of Aix is a speciality having nothing in common with that practised in the capital towns, which has become too much in general vogue.

It would be useless to dwell at length on the advantages of massage. In England many physicians, including Drs. Murrel, Thomas Dowse Ryley, and others; in France, Norstrou, Gilles, and others, have written exhaustively on the subject. On all points I agree with them, but I regret that massage, so useful in many cases, should fall into the hands of charlatans of both sexes, who trade on the credulity of the public without considering the counter-indications which are frequently present. At Aix the advantage consists in the safety of a system never employed without the authorisation and direction of the attendant doctor.

I can quote the case of a young American lady suffering from tuberculosis, who fell a victim to peritonitis precipitated by an injudicious massage. Also the case of a colonel, who expired suddenly after an attack of angina pectoris, the result of a fantastic massage not indicated by any doctor.

An article by Dr. Grainger Stewart, of Edinburgh, who followed the course some years ago, describes so thoroughly the process, whilst proving his appreciation of the administration and results, that a brief extract is appended:—

"It is astonishing with what skill, what patience, tenderness, and firmness the shampooing and passive movements are performed. When every joint has been moved to the utmost extent possible, the patient is made to stand, while from a distance a powerful stream of water is propelled upon the different limbs, especially about the articulations chiefly affected. According to circumstances, he may have warm water to the last, or he may have a cool or even a cold douche, or perhaps a good cold shower-bath, which in Aix is known, whether in compliment to our climate or for some better reason, as the *Douche Ecossaise*. When the bath is over, the patient is rapidly dried, wrapped in flannel sheets and blankets, and is carried back to his hotel in the curious sedan-chair. Having reached his apartment, he is lifted into bed, still swathed like a mummy, is covered up with additional blankets and a quilt, and left to perspire for a longer or shorter period. After twenty minutes or half an hour, he is carefully rubbed down by an attendant, who had accompanied him to the bath. If the case be less severe or the patient feeble, he may not be subjected to this post-balneal bed perspiration.

"As an occasional variety, instead of having the douche, the patient is sent to have a **steam bath** —the Berthollet, as it is termed in Aix. He enters an apartment which contains a curious wooden box, with a round hole in its movable lid. After undressing, he steps into the wooden box, and finds that he is shut in all except the head, the round hole being occupied by his neck. Immediately a valve on the level of the floor is opened, the hot vapour rises about him, and he soon begins to perspire freely. The perspiration running from his brow, trickles down his face. Presently he feels the streams flowing down his sides and his legs, and very speedily a feeling of oppression and debility comes on, and after ten or twenty minutes the bath is opened up, the patient is carefully dried and removed to his hotel.

"Sometimes on the same day as a douche, and sometimes as the sole treatment, the patient gets a **local vapour bath.** By ingenious contrivances the bathman is enabled to steam one arm or one leg. Speedily the limb begins to perspire, and the parts become soft and comparatively flexible. Perspiration occurs all over the body, especially in those who have been undergoing other forms of treatment, and so great care requires to be taken to prevent a chill.

But the patients often sit and read while one arm or leg is exposed to the vapour. When the parts have been thoroughly softened, manipulation, shampooing, and passive movements of joints are carefully carried out, just as after the douche, but only confined to the one limb.

"On certain days the patient is sent to the spacious and comfortable **swimming baths**, and there he is allowed to disport himself for a longer or shorter time, practising amid the somewhat warm water active movements of his limbs. As his joints relax, he may find that he can cross the bath in ten strokes instead of the fifteen that were originally necessary, and that he can continue the exercise much longer than he could at first. When his swim is ended—and swimming in warm water is rather enervating—he may have a cold douche or not according to the directions of the doctor. He is rapidly dried, and if well enough, is directed to walk about smartly in the gardens, which are close to the establishment."

Baths.—Few establishments can provide so many baths at different degrees of temperature and mineralisation, thanks to the large quantity of water at our disposal. Natural water is added to the bath to diminish the temperature and mineral strength. By the addition of some litres of Challes water, baths extremely rich in sulphurous principle are obtained; if tepid, they preserve the stimulating property of their mineral constituents.

Inhalations.—Two large inhaling rooms, one possessing a perfect system of humage, have been fitted up. Here the patient inhales the sulphurous vapour in various ways, according to the mode prescribed. The inhaling apparatus delivers the steam without necessitating any inspiratory effort.

Spray.—Thermal water, atomised in spray, may be directed on every part of the body; and in this form is especially employed in throat and nose diseases, in affections of the eyes and face, and where there is relaxation of the mucous membrane. Water thus pulverised loses no portion of its medicinal power, and penetrates most readily into every fold of the skin and mucous membrane.

GENERAL AND THERAPEUTIC ACTION OF THE WATER:

THERMAL FEVER (POUSSÉE).

The action of these waters affects essentially the excretions of the skin and of the kidneys, and influences the respiration and circulation. Hence all the physiological processes are increased; the appetite is strengthened, the muscular tone developed, and the menstrual functions are regulated.

The **douche and vapour baths** are especially exciting, for the surrounding air is raised to a temperature of 105 or 110 degrees, and charged with watery and hydro-sulphurous vapour which sensibly diminish the oxygen, and thus the rapidity of the respiration, and consequently of the circulation is increased.

Many years ago, a distinguished physician of

Lyons made numerous physiological observations on the results of his treatment at Aix, and these are confirmed by personal experience among my hospital patients. The heat of the body rises during the douches two or three degrees, and that of the arterial pulsations from 70 to 100 degrees.

Some sort of artificial fever is produced, which naturally finds vent in abundant perspiration; at the same time the organic matters contained in the urine are excreted in large quantities; the shock from two or three douches of hot water for several minutes, accompanied by shampooing, adds to the general excitement. The ensuing disturbance of the system necessitates great precaution. Any imprudence or exaggeration may develop pathological phenomena; but by prudently and steadily pursuing the stimulating treatment for three or four weeks the happiest results are attained.

Further on I will point out the counter-indications of the course. The **thermal fever** or slight fever, sometimes complicated by gastric symptoms, yields in most cases to rest and diet.

When the tissues of the skin are very delicate, a slight irritation occurs; but never in my long practice have I observed the *pousseé* (so called) in a severe form, or complicated with fever. In addition to their stimulating properties, sulphurous thermal waters produce a special effect commonly ascribed to electricity, but which in point of fact is neither understood nor defined, and which never exists in sulphur waters artificially compounded.

DURATION OF THE TREATMENT.

The length of the course cannot be arbitrarily defined; it varies according to the age, sex, and constitution of the patient, and the nature and stage of the illness. Evidently nervous women, children, old men, and those long invalided, are unable to support so long or severe a treatment as adults seeking relief from rheumatism, sciatica, or laryngitis.

Before starting, the habitual medical adviser should be consulted; afterwards, the directions of the local doctor who watches the progress and powers of endurance of each individual must be scrupulously observed. It is often found advisable temporarily to suspend the treatment. When taken under satisfactory conditions, the ordinary course lasts about twenty-five days, with brief intervals of rest.

An after cure of a week's fresh air in some adjacent mountain retreat before returning to the daily routine of life—a system more in vogue in Germany than in France—contributes greatly to the success of the treatment.

I often send my patients to St. Gervais, to Chamonix, or to Brides-les-Bains, according to the case. Shortly we shall possess, in our mountain of the Revard, a health resort of perfect salubrity, within forty minutes, railway of Aix.

BEST SEASON FOR THE COURSE.

The moderation and regularity of the climate enable the treatment to be followed during six months of the year; however, I do not advise patients to arrive before the 25th of April or after the 15th of October. The most suitable season dates from the 1st of May to the middle of October; for the action on the skin opens the pores and renders the patient unusually susceptible to changes of temperature. As the seasons vary from year to year, it is prudent before fixing the date of departure to write to the local doctor. The months of July and August are very hot, but English visitors by avoiding the close rooms of the Casinos during the afternoon, and seeking fresh air and oxygen on the neighbouring heights, will find them quite endurable. In support of this view I may quote the opinion of Dr. Prosser James, a regular visitor to Aix, and author of an excellent work on the mineral waters of Europe :—" It is often said that English people should avoid Aix in August, but personally I have not found it too hot in that month; and in those severe rheumatic cases which are sensitive to cold, there are several reasons for choosing that time for a course of baths."

DIET DURING THE COURSE.

Though frequently consulted on this point, I cannot lay down general laws; the diet of each patient is regulated according to his special malady, constitution, and ordinary habits, and in the first place by the advice of his own doctor. In the hotels the food is simple, but of excellent quality, and thoroughly adapted to the course; no kind of venison or of pork is provided. It is best to eat sparingly of brown meats, and fresh water fish in preference to sea fish; to vary the diet as much as possible; and to make no change in the customary beverages. When wine is not drunk, the local doctor occasionally prescribes some form of tonic. Constipation must be combated, particularly in gouty and rheumatic cases. Constant exercise in the open air is essential The damp atmosphere of the spring and autumn evenings, late hours, and every kind of excess, are prejudicial.

RHEUMATIC DISEASES.

Of all known diseases, rheumatism has been most used and abused for establishing the merits of different spas, but to our waters we unhesitatingly ascribe the specific quality of healing rheumatism, through the process of external and internal absorption of sulphuretted hydrogen, which eliminates the generating poison through the urine and the pores of the skin, as proved by the experience of centuries and confirmed by recent scientific observations.

Dr. Hermann Weber expresses himself on this subject with a certain amount of scepticism easy to understand and almost to adopt. "There is no bath which would not be recommended as a remedy for rheumatism, and this disease, moreover, belongs to those which possess hundreds of popular cures, and to which the artificial and natural remedies of the enthusiast and impostor are applied with especial predilection. Many a champion of quackery has become rich through rheumatic chains and anti-rheumatic balsams; and among the cases recommended for Hoff's malt extract and Lampe's cure with simples, rheumatism stands at the head. The word rheumatism, although so familiar to the physician and to the ignorant, does not designate an idea forming a scientific or even practical category for individual cases and groups of cases. Of all that we call rheumatic, nothing but acute rheumatism of the joints presents an actual and exact idea of disease; and everything else bearing this name forms a chaos of different and partially undefined conditions." * In fact, every pain, from the slightest twinge to the most excruciating form of neuralgia, is termed indiscriminately rheumatism. We treat with certain success—

1. Torpid rheumatism, of a lymphatic or scrofulous nature, in varied constitutions—by douches, shampooing, and swathing.

2. Rheumatism, complicated by neuralgia, with neuropathic constitutions and visceral metastasis—by simple baths or swimming baths, slight douches of about the temperature of the body, or even lower (atony, anæmia).

3. Rheumatism, complicated by skin troubles—by the more highly mineralised waters of Marlioz and Challes.

In the two latter cases the patient, by gentle exercise, stimulates the reaction due to the exciting draught of mineral water.

* "Curative Effects of Baths and Waters." By H. Weber. 1875.

Later on, the beneficial effect of the treatment in joint diseases, based upon rheumatism, will be pointed out.

Some years ago I presented to the College of Surgeons in Paris some notes on a case of traumatic and rheumatic tetanus which had been cured by our course. Unfortunately, they are too voluminous for the limits of these pages. To me, tetanus represents an exaggeration of those phenomena which constitute rheumatism.

Cases derive relief at Aix when the pain manifests itself in the envelopes of the nerves, sheaths of the muscles, tendons, fibrous textures of the joints, or in the periosteum. This treatment is also efficacious in lumbago, pleurodynia, and gonorrhœal rheumatism.

"The most frequent cause of this affection is the previous existence of gonorrhœa, but sometimes a history of leucorrhœa only can be traced; hence the propriety of the term 'genital rheumatism.' In sub-acute and chronic rheumatic synovitis the knee-joint is generally affected, and the form of the swelling, corresponding to the synovial sac, is diagnostic; fluctuation is distinct; there is some increase of heat and a little pain, but no redness of the skin, and no disposition to suppuration or ulceration."[*]

During the last few years I have observed many cases of complete cure of these cases. "Gonorrhœal rheumatism occurs in connection with gonorrhœa. It is not accompanied by the same amount of febrile disturbance as acute or even sub-acute rheumatism. It affects

[*] "Lettsomian Lectures." By William Adams, F.R.C.S. 1869.

fewer joints, has a special preference for the knee, and does not show the same tendency to shift about. Acid perspirations do not occur, and it does not tend to affect the heart."* Dr. Maclagan's lucid summary proves that this form of rheumatism admits of a stringent course, without incurring risk of metastasis or cardiac repercussion.

Dr. Maclagan, who first applied salicine to the cure of rheumatism, approves also of sulphurous water in rheumatic cases. "The treatment of some of the sulphurous baths of France and Germany holds out the best prospect of relief."†

Experience proves that our waters not only cure rheumatic manifestations, but combat the hereditary rheumatic diathesis; and after several courses impart to the constitution the power of resisting the encroachment of the rheumatic poison.

Our treatment, when pursued vigorously, is slightly debilitating. Thus, Sir Alfred Garrod, a great authority in this matter (quoted by Besnier, p. 709, "Dictionnaire des Sciences Médicales"), declares, that according to his recent observations of the effects of mineral waters, the results of hydro-thermal treatment are frequently unsatisfactory, either because they are not adapted to the individual case, or because they have been injudiciously and very vigorously applied. The waters of Vichy, Carlsbad, and Wiesbaden are hurtful when not administered with great caution; those of Aix in Savoy, of Marlioz and of Challes, have

* "On Rheumatism." By Dr. Maclagan. Pickering, 1881. P. 6.
† *Ibid.* p. 223.

furnished more satisfactory results, and it has always proved beneficial to supplement this course with a tonic treatment at Schwalbach Spa, or St. Moritz.

Professor Stewart's note will be read with interest:—

"The treatment at Aix is of extraordinary value in various rheumatic conditions. *First,* it is of great service in the way of removing the thickness and stiffness which so often remain after attacks of acute rheumatism—a stiffness due partly to changes within the joint, but mainly to thickening of the fibrous tissues around the articulation. *Second,* in cases of chronic rheumatism, where a slow inflammatory action is going on in and around the joints, it suffices both to remove inflammatory products and to diminish the tendency to rheumatic inflammation. *Third,* in rheumatic affections of the muscles, fasciæ, and nerve sheaths, it affords in many cases the most decided and speedy relief. *Fourth,* in the wasting of muscles, which so often occurs in connection with rheumatic processes, the manipulation and shampooing, along with the electrical stimulation which the doctors superadd, generally prove distinctly serviceable ; and *fifth,* on the occurrence of slight rheumatic threatenings it appears that the use of the Berthollet, or vapour bath, often suffices to prevent the further development of the disease."*

The proper temperature of baths and douches, and the density of vapour, depend on the state and suscep-

---

* *Edinburgh Clinical Journal,* October 29, 1883.

tibility of the individual. Some patients cannot support simple baths, and are greatly benefited by douches and vapour baths : others can be treated only by simple baths. Thus, our course necessitates continual medical supervision, based especially on frequent analysis of the urine. The length of the shampooing and sweating processes, and the duration of the entire course, varies according to the torpid or inflammatory symptoms of the case.

RHEUMATOID ARTHRITIS.

Rheumatic Gout, Chronic Rheumatic Arthritis, Nodosity of the Joints, Nodosity of Heberden, Arthritis Deformans—is not a blend of gout and rheumatism as commonly supposed.* Sir Alfred Garrod quotes Fuller to this effect :—" The disease should not be regarded as of a hybrid character, or, in other words, made up in part of rheumatism, in part of gout." Rheumatic or gouty persons may develop rheumatic gout ; but even here the arthritis cannot be regarded as causally related to either. " It appears to result from a peculiar form of mal-nutrition of the tissues of the joints, being an inflammation accompanied with defective powers ; but there is no evidence, upon which any reliance can be placed, to show that it depends either upon the presence of any morbid principle, or upon a weakened condition of the vessels or structures of the affected parts."† It is a disease of

---

* " Gout in its Protean Aspects." By J. M. Fothergill, M.D. Lewis. 1883. Page 242.

† Garrod " On Gout."

debility—that may be affirmed—free from any tendency to affect the heart, or to induce any kidney change.

Rheumatic gout, then, is a disease *sui generis*, in which there is general atony and a depraved nutrition of the ends of bones, of the epiphyses indeed, to speak, broadly, and dyspepsia occurs incidentally. Fuller, who gives great attention to it, says: " Its earliest attacks are usually seen in girls whose uterine functions are suspended or ill-performed ; " while "it invades the stiffening articulations of the woman who has arrived at that time of life which is marked by the cessation of the monthly periods ; it shows itself during the state of debility which follows a miscarriage, or a difficult and protracted labour, more especially when the labour has been accompanied by flooding."

Many years ago I translated a very interesting pamphlet by Dr. Ord, on " Hysterical Rheumatism of Women at a Critical Period, or in connection with certain Uterine Derangements ; " since then many such cases have come under my notice, which I, in common with the physician of St. Thomas's Hospital, denominate " Hysterical Rheumatism." The diagnosis of this affection becomes easy after a certain experience. When the disease is advanced, no certain improvement can be obtained in the bones (epiphyses); but progress of the malady may be delayed, the periarticular swelling diminished, and a certain elasticity restored to the articulations ; thus, I always warn the patient that we can only hope to stem the progress of the disease.

Local sulphurous vapour baths are peculiarly successful if taken twice daily with two glasses of Challes water tinged with iodide of potassium, or of lithia, according to the case. A new stimulus is imparted to the system, the organic functions receive an impulse not obtainable in any other way; and remedies which have proved unavailable at home shortly become active agents for good. Fuller thus explains the action of the water and of our treatment. "This purely essential affection is due to a specific poison on which our waters act by an eliminating process through the two organs of secretion. We occasionally observe, in women of a certain age, arthritis of the finger-joints or inflammation of the hip and shoulder.

"In chronic rheumatism our attention is frequently called to the deformity of the great toe, consisting of a deviation when the soles of the two feet touch each other. This is produced by a proliferant osteitis on a level with the internal surface of the metatarso-phalangeal joint of each toe. Sometimes the skin is the seat of chronic inflammation, and of pain, especially under the pressure of the boot.

"Chronic forms generally first present themselves in bony articular or pre-articular lesions; cases in which the alterations are limited to the soft parts of the joints appear scarce, especially in France.

"Rheumatism producing deformity commences in the small joints. The nodosities of Heberden are regarded as the expression of the slightest and attenuated form of the diathesis. The outward deviation of the great toes precedes these. I have never encountered these nodosities where the malformation is not very evident. The fact of the great toe being generally the seat of the diathesis is accounted for, not only by the pressure of the boot, but because the whole weight of the body is thrown on those joints.

"Dr. Besnier, in his article on 'Rheumatism' in the *Medical Dictionary*, attributes to functional abuse a powerful agency in these local articular rheumatic affections.

"Professor Verneuil, in one of his clinical lectures in 1883 (*Semaine Médicale*), indicates this deviation as 'a certain sign of the arthritic diathesis.' I have met with this malformation of the great toes in children five or six years old, and in young girls, who show no other rheumatic symptoms. The pressure of the boot being sometimes considered the sole or chief cause, I naturally inquire: How then account for these affections in children? Certain cases of neuralgia, myalgia, and even of arthritis, which have come under my notice, without offering this first impression of the rheumatic diathesis, have been attributed, after more or less careful observation, to albuminuria, glycosuria, lead poisoning, syphilis, or scrofula.

"This malformation of the great toe is not uniform, one toe being almost always more crooked than the other, thus proving the predominance of the affection on one side of the body. This peculiarity gives to the malady an hemiplegic appearance, in respect to its origin. The predominance of rheumatism occurs sometimes on the right, sometimes on the left of the body; but apparently more often on the left. The hemiplegic appearance of deforming rheumatism is not limited to articular lesions; it is applicable also to abarticular manifestations.

"In the frequent cases of complications in the bronchial regions, the wheezings at the *base* appear more common on the side most affected by rheumatism. In one of the last *séances* of the Berlin Medical Society, Professor Virchow observed, 'Care must be taken not to confound deforming arthritis with *uratic* arthritis: these two forms rarely co-exist in the same subject. I have only met with one case in which I have proved the presence of uratic deposits distinct from nodosities.' The morbid process of the gout and that of deforming rheumatism being dissimilar, the alteration in the form of the great toes must not be considered identical in the two cases.

"This difference in the material changes of the regions might prove in doubtful cases a valuable sign of the differential diagnosis. The uratic infiltration which characterises gout, and the proliferant osteitis which characterises deforming rheumatism, cannot leave the same impression on the parts affected. I have seen a great many rheumatic cases, but have less experience in gouty cases. It appears to me that in gouty cases the great toes touched, or were but slightly separated from, each other. I have clearly observed this in a gentleman who in three years had had four acute attacks, of which two were very severe. In deforming rheumatism, the exostosis (Virchow) which occupies the internal surface of the metatarso-phalangeal joint of the first toes bends the great toes outwards. The accumulation of matter occurs in the entire joint, and increases its size; in rheumatism the osteitis is produced inside, and is partial and essentially deforming.

"I arrive at the following conclusions :—

"1. In the bony forms of chronic rheumatism a special deformity of the great toes exists, the deviation occurring outwards, which is a more evident sign of the diathesis than any other articular localisation.

"2. Any other peri-articular manifestation, with a gouty appearance, not coinciding with this deformity must be looked on as doubtful, and the cause will probably be traced to some other existing malady or some other diathesis.

"3. This deformity is not uniform; the preponderance on one side extends to the corresponding half of the body. It also imparts to rheumatism a hemiplegic aspect as to the local seat.

"4. The change of form differs from that produced by gout, the two morbid processes being essentially dissimilar.

"5. Might the pathological distortion of the great toes furnish the element of a comparative diagnosis between deforming rheumatism and gout?"

GOUT.

THE etiology of this disease is well known, thanks to the researches of Tennant, Forbes, Murray, Wollaston, Hood, Fuller, and especially of Garrod and Charcot.

The origin is traced to the presence of uric acid in the blood, frequently affecting the joints and muscles.

For many years a prejudice, still extant, precluded the use of sulphurous waters in these cases. This is easily accounted for by the dread of our course entertained by certain doctors in the beginning of the present century, when all patients, irrespective of age and constitution, underwent the entire treatment—the hot bath, steam bath, and swathing—whilst now a third of our patients only take the baths and douches at a temperature of $96°$ Fahr., and are carried back to their beds without the enforced perspirations—too exciting in certain cases.

Gout, better understood than rheumatism, is subject to acute crises, frequently excited by the former mode of treatment. Even at present gouty as well as rheumatic and sciatic crises may be evoked by our course, but these crises are generally advantageous, and the course is only suspended until acute pain subsides. It required all the influence of a renowned physician such as Sir Alfred Garrod to restore to Aix its reputation as a remedy in cases of gout; which, judging from the benefit derived by the numerous patients to whom he had prescribed our waters, he was in a position to appreciate fully. In page 427 of his last edition (1876), Sir Alfred Garrod expresses a positive opinion in favour of our course, when gout manifests itself as cutaneous eruptions such as psoriasis and eczema, or as swelling and fulness of the joints, or as forms of gout not connected with portal congestion, where the influence of ordinary thermal waters is equally efficacious. There would be nothing to add to the emphatic assertion of this great authority, were I not in a position to prove by long personal experience among English patients the eliminating effects of our course, whereby the frequency and intensity of gouty crises are diminished, although at the beginning of the treatment a slight manifestation may be produced. Strict diet is invariably prescribed, and careful medical superintendence must be exercised, especially in respect to the urine. In addition to the therapeutic course there is the mechanical treatment which I apply by means of continuous currents in cases of gouty joint diseases and muscular atrophy.

Short and frequent vapour baths, followed by gentle shampooing, exercise a specific action on those articular deformities which gradually lead to ankyloses; absorption of the effusion in the bursæ is thus effected, and stiffness gradually disappears.

I have even seen cases where the movements of the joints were facilitated by the precipitation of chalkstones.

We have equal success in atrophy of the muscles proceeding from gout, which is the more serious because the arthritis is far more severe, as Sir James Paget asserts in his "Lessons on Clinical Surgery." This muscular atrophy, which does not resemble the muscular atrophy in the arthritis described by Charcot, results exclusively from gouty and rheumatic lesions. Baths, douches and shampooing, supplemented by electric currents, are highly beneficial.

As I might be supposed to be too partial in my views regarding the treatment at Aix, I cannot do better than quote the greatest authority on gout and rheumatism from the very interesting article published by Sir Alfred Garrod in the *Lancet* of May 4, 1889 :

"I have already said that the soothing character of the climate and waters renders Aix a suitable place for the treatment of diseases of the respiratory organs, but I do not propose in this communication to speak further on the subject of these affections, but to dwell chiefly on the therapeutic value of the Aix course on patients suffering from different articular, muscular, skin, and some nerve diseases.

"In the first place, we will select three forms of disease—rheumatoid arthritis (rheumatic gout), gout, and true articular rheumatism, the last of which in its acute form is commonly called rheumatic fever.

"It is most essential that we should make a clear diagnosis of these three maladies, for although they may, in some of their stages, very closely resemble each other as far as external appearances are concerned, yet in reality, with the exception that all of them are articular diseases and often attack the same joints, they have nothing in common, and their treatment, both medicinal and dietetic, requires to be very different. The course of Aix, although often useful in all three diseases, is notably far more valuable in one than the other two.

"Under these circumstances I shall make no apology for devoting a short time to the endeavour to separate these diseases, which in practice are constantly mistaken for each other, an error of diagnosis frequently leading to much and often irreparable mischief.

"For this purpose it may be well to picture to ourselves three cases, one of each disease, and by a careful search into their histories and minute particulars, try to come to a correct estimate of the true character of each.

"The first case shall be a young woman or man, say, from twenty to thirty years old, having a distinct swelling of one knee, also with slight increase of temperature of these parts, but not very acute pain. There is little constitutional disturbance, but distinct debility, possibly more or less anæmia.

"The second case shall be a young man, possibly woman, of about the same age, with the same amount of joint affection—say, in one foot and ankle and one knee; no great pain, but some tenderness and increased warmth of the implicated parts. In this case also the constitutional disturbance may be slight—in fact, scarcely appreciable; and very little debility may be present.

"The third case shall be a young woman or man exhibiting the same amount of articular affection, with the same joints implicated; the constitutional disturbance at the time very little, but usually more or less accompanying debility.

"How are we to diagnose correctly these three cases which in external appearance resemble each other so closely? They may represent certain stages of diseases which are essentially different from each in their pathology, and which in other stages could be more readily separated.

"Of the three cases above alluded to, the first is one of rheumatoid arthritis (rheumatic gout); the second, one of true gout; the third, one of genuine rheumatism. Let us now look more minutely into these case.

"*Case I.*—In the history of the first case we shall probably find that some severe depressing cause has been in operation before the development of disease. The cause may have been either mental or physical; there may have been some well-marked heredity on one or both sides; the disease has probably come on very insidiously. In some cases, however, the invasion may be rapid; it is almost certain to have been progressive; the disease has travelled from joint to joint, but the invasion of a new joint is not accompanied with amelioration of those previously affected. If we inquire, it is not unlikely that we shall find that there is or has been some pain and stiffness of the back of the neck and jaw, not looked upon by the patient as connected with the other symptoms. There is more or less pitting of the swollen parts, but no evidence of cardiac or kidney disease arising in the course of the joint affection.

"As to the question of any connection between alcohol and rheumatoid arthritis, I can positively say that it occurs not unfrequently in patients who have never in their lives taken alcohol, and I have seen some instances in which even the parents also had been complete abstainers.

"As to the nature of rheumatoid arthritis, I have given elsewhere my opinion to the effect that it results from a peculiar form of mal-nutrition of the different tissues of the joints, accompanied with defective power. There is no evidence to show that the symptoms depend upon presence of any morbid principle in the system. Lastly, there is much to show, in its etiology and the distribution of the affected joints, that it is intimately connected with the nervous system. Two papers in the 'Medico-Chirurgical Transactions' of last year go far to prove this.

"If in such a case as we have described the blood is examined, no uric acid is found; and if an opportunity occurs of observing the state of the joints, there are no urate deposits seen, which is invariably the case in true gout; but in place of such we find, from the first, evidence of ulceration of the cartilages, which goes on increasing till the whole of this structure is removed, and complete disorganisation of the joints ensues. Very many patients who have during the past few months exhibited little or no more than we have pictured in this our first case have, within a year or so, become hopelessly crippled and practically bedridden. I feel quite sure that in many such cases an erroneous treatment has much to answer for: errors in diet, errors in medicine, and errors in bath treatment, have frequently united in bringing about this lamentable condition.

"*Aix-les-Bains Treatment in Rheumatoid Arthritis.* —It is in the treatment of cases belonging to this class of maladies that the course at Aix-les-Bains is peculiarly successful, not only in ameliorating at the time the condition of the patient, but in removing the tendency to the disease from the system.

"The effect of the Aix waters is by no means confined to the period of their administration, but the improvement often continues for several months; and now and then a patient, who may be disappointed at the termination of the course, after a short time has full reason to feel satisfied. I have seen numerous cases of rheumatoid arthritis most signally benefited by the treatment, and no relapse has occurred after several years.

"To effect this amelioration in cases of rheumatoid arthritis several influences are at work; first, that of the hot sulphur mineral douche; secondly, the massage accompanying the douche; thirdly, the internal exhibition of the sulphur water, whether it be the Aix water or that from Marlioz or Challes.

"These, together with the climatic influence, produce an effect on the system which is most valuable. My own experience leads me to think the Aix course, if properly pursued, is not debilitating, or, if a feeling of relaxation is produced at first, it is by no means permanent. I have been frequently asked why Turkish baths should not prove equally as beneficial as the Aix treatment in cases of rheumatoid arthritis. My own experience enables me to say with confidence that, however useful they may be in other forms of articular disease, in this special affection their use is certainly not indicated, but strongly to be condemned, and I have seen in many cases much injury done by them: in one case—that of a young woman, to whom they were cruelly exhibited for thirteen weeks and often twice in the day—hopeless crippling was induced, and almost every joint in the body became completely ankylosed, so that she could neither stand nor sit in a chair without being propped up with pillows. Patients and their friends are apt to be deceived by the fact that when taking the Turkish bath, and even for a short time afterwards, the pains are often relieved, but no permanent benefit is obtained; the subsequent mischief is in proportion to the debility induced.

"I do not wish for a moment to be supposed to consider the Aix course as a certain cure; there are many cases in which the nutrition of the body is so hopelessly undermined that no treatment has a chance of proving effectual, either in removing the symptoms or checking the progress of the disease; and, moreover, we must not trust to mineral water alone, for I am confident that medicinal and dietetic treatment, continued over a lengthened time, is most essential. We must not lose sight of the fact that rheumatoid arthritis is essentially a slow but progressive disease, and a few weeks' treatment cannot be relied on to produce a permanent cure; of this, however, we may feel sure, that everything which improves the general state of health has at the same time a tendency to lessen the disease.

"*Case II.*—We will now investigate our second case, that of the young man suffering from a slight attack of true gout. We shall find a history very different from that of the first case. There is a great probability that he will have inherited gout from his parents or grandparents, on one or other side, as he has been attacked with the disease so early in life; and there is an equal probability that he has indulged somewhat freely in those alcoholic fluids which cause gout. We shall also find that he had several previous attacks, the first most likely confined to the ball of the great toe; that the disease is periodic, but not continuous, as in the case of rheumatoid arthritis; there need not have been any case of rheumatoid arthritis; there need not have been any previous depressing causes in operation, but eczema, cramps, and dyspepsia may have preceded and accompanied the articular affection. In such cases we do not find cardiac affection as part of the disease, but in those of mature years valvular affection may be present as a result of the long-continued use, or rather abuse, of alcohol, and there is a great tendency to kidney disease. An examination of the blood in such a case reveals the presence of much uric acid, and evidence of the deposit of urate of soda is often seen on the helix of the ears and elsewhere: and if at any time an opportunity is afforded of an examination of the joints, streaks and patches of urate of soda are seen on the surface of the cartilages and ligaments.

"*Aix-les-Bains Treatment in Gout.*— When the disease is any way recent or acute, the Aix treatment should be avoided; for if the local inflammation were checked by the douches and massage it is sure to be developed in some other part; for there is always a morbid state of the blood in gout, different from what exists in rheumatoid arthritis, and the morbid elements must be eliminated somehow or other. Hence mischief, rather than good, may ensue from the Aix treatment, which has no very great power of freeing the blood—at any rate, not in sthenic states of the system. The case is different in asthenic and chronic forms of gout; in these there is usually a defective power of the nervous system and a weakened state of the circulation. When there are passive or indolent swellings of previously inflamed parts; when there is much eczema; when there is a threatening of kidney disease in which a free action of the skin proves of much service; when massage is indicated—then in all such cases the value of the Aix course is great and undoubted, and it often lengthens considerably the interval between the attacks of gout.

"*Case III.*—The examination of the third case will readily reveal the nature of disease under which the patient is suffering. In all probability there have been one or more previous and acute attacks of long duration; that many joints have been implicated; and that in these attacks there has been much febrile disturbance and excessive perspiration. Not unlikely there has been pericarditis or endocarditis, which has left its marks. We shall also find that cold has been the immediate cause of the illnesses, favoured by a previously weakened condition of the body; that the disease has not been continuously progressive, but that there have been intervals of good health, if the heart has not been a cause of trouble. The blood in such a case exhibits no evidence of the presence of uric acid, perhaps it may be less alkaline than in health; and if an opportunity occurs of examining the joints, although we may find evidence remaining of more or less inflammatory action, there are no deposits on the cartilages or ligaments of urate of soda as in gout, and no evidence of ulceration of the cartilages as in rheumatoid arthritis. This third case is one of a subacute form of true rheumatism, which is so frequently met with in the intervals of the acute attacks of this disease.

"*Value of the Aix Course in Articular Rheumatism.*—No one would think of submitting a patient to the Aix treatment in articular rheumatism if there is any amount of febrile disturbance lingering about, or if any recent cardiac disease is present; and, as a rule, the joints after even an acute attack are not left in a very disabled condition. I once saw a comparatively young woman who had experienced twelve attacks of acute rheumatism; yet after recovery from the last attack she exhibited no very detectable articular changes, except a certain amount of relaxation of the joints. The Aix treatment therefore is not often required in such cases, but should any stiffness remain it would doubtless be of value. The question I am frequently asked is, whether after recovery from an acute attack a course at Aix would have any influence in preventing the return of the disease? I am not able to speak very confidently on this point from clinical experience, and one must see on reflection that the difficulty of arriving at a correct conclusion must be very great; still this much may be said, that the Aix course acts as a general alterative to the body at large, and certainly has a beneficial effect on the nervous system, improves the general health, and has therefore in all probability a protecting influence against recurrences. Some physicians have asserted that valvular murmurs often disappear; this subject, however, requires more clinical study before it can be looked upon as established.

"*Muscular Rheumatism, Sciatica, and other Neuralgic Affections.*—In chronic muscular stiffness there can be do doubt of the benefit frequently derived from the hot sulphur douche, and often from the massage; and the effect of these appliances is increased by the drinking of some sulphurous waters, in the same way as we have seen them benefited by the long-continued use of small doses of sulphur. Neuralgic affections, as of the sciatic nerve, and also of different branches of the brachial plexus of nerves, when they have become chronic, also receive much benefit from the Aix course."

NEURALGIA.

The forms most frequently treated with success at Aix are brachialgia, pleurodynia, and sciatica. Patients suffering in the hip, the leg, or even from lumbago, indiscriminately designate their complaint as sciatica, though in damp countries articular and peri-articular pains, closely resembling sciatica, often result from an actual arthritis of the hip.

NEURALGIA.

I insist on this point, because the treatment differs according to the malady; whilst real sciatica derives great benefit from a vigorous course, its counterfeits are intensified under a similar treatment. Great care must be taken to ascertain whether the pain arises from rheumatism, gout, syphilis, exposure to cold and wet, or simply from over-fatigue. When due to syphilis, sciatica can only be cured by the specific treatment, but such cases are rare. Professor Fournier observes that neuralgia rarely affects the entire nerve, and never descends below the knee.

I have observed three cases of partial sciatica, proved to be of syphilitic origin. When there is evidence of the specific element, the waters of Aix produce the double effect of combating the local manifestation and facilitating the endurance of the required form of treatment. If pain is acute, I consider the simple bath preferable to the vapour bath, as it acts sufficiently on the skin without over-exciting the nervous system; but where neuralgia is torpid and atonic, the large douche becomes requisite. When the patients are in a very debilitated state, plain swimming baths succeed very well. Every year recovery ensues in numerous cases of sciatica from a course of sulphur baths, in which the mineral element excites and regulates the functions of the skin. Long-standing pain, where atrophy has produced immobility of the limb, necessitates a prolonged treatment. First the element of pain, and then the ravages caused by the malady, must be dealt with. Douches and shampooing constitute the final cure.

In neuralgia, as well as rheumatism and other similar affections, an increase of pain frequently occurs during the first days of the treatment, and may last for some time. Almost as many cases of neuralgia alone are cured at Aix as of the entire category of other forms of rheumatism. Nine-tenths of the cases thus treated, if not completely cured, experience considerable relief during the first season, and success is certain after a second or third season.

Case of Gouty Sciatica—Complete Cure.*—" Mr. ———, nearly thirty years of age, of a bilious, sanguine temperament, of robust constitution, subject to flying gout, wearied by suffering for many days from a sharp pain in the region of the right instep, applied upon this part a nearly cold linseed-meal poultice. Some hours later the pain in the foot had vanished, but was succeeded by acute pain in the posterior region of the thigh, in the course of the sciatic nerve. In vain it was tried to bring back the inflammation to its primitive seat—all means failed. Leeches were applied in the track of the nerve, then flying blisters, and, lastly, the actual cautery. No progress was made. The invalid walked with much pain, supported by crutches. The pains in the track of the nerve did not cease; they were specially very sharp on movement of the limb. The saline waters of . . . . were recommended. The patient went to the place, drank the waters, took the baths and the douches. No benefit accrued after a month's residence. The state of matters remained the same until

* Professor Tnissac (Montpellier).

the commencement of the summer of the following year. He then went to the sulphurous spa of Aix, in Savoy, a cripple, walking by the aid of crutches. He there made use of the waters internally and externally by baths and douches. A speedy amelioration showed itself; he made rapid progress, and after five weeks' treatment threw away his crutches, and recovered the complete use of his limb. Twelve years have since elapsed, and the sciatica has not returned. Mr. ———— is not altogether cured of the gouty diathesis, for now and again he suffers from intestinal colic, cramps of the stomach, sometimes from palpitations of the heart, with uneasiness in the region of this organ, sometimes from pains in the joints or in the lumbar region; but these various symptoms, although very inconvenient, do not continue for many days, as they succumb speedily to well-directed remedies. The health, with the exception of these attacks, is very good."

DISEASES OF THE ORGANS OF RESPIRATION.

RHINITIS—OZÆNA.

The waters of Marlioz and Challes are highly efficacious in cases of chronic inflammation of the nostrils. We frequently meet with gouty subjects, and especially strumous children with swollen pituitary membrane, emitting a fœtid odour from muco-purulent discharge, and caries or necrosis of spongy bones in severe syphilitic and scrofulous cases. The nasal douches of Marlioz, the inhalation of natural sulphurous steam, spray administered locally, swimming baths, with the internal use of Challes waters, closely resembling in effect cod-liver oil, produce a general invigorating influence.

PHARYNGITIS.

(Catarrh of the Pharynx—Sore Throat.)

Granular Pharyngitis (*Clergyman's Sore Throat*).

Herpes of the Pharynx (*Rheumatic and Gouty Sore Throat*).

We never treat acute pharyngitis, but often meet with patients in whom acute inflammation has left great weakness of the mucous membrane, with the predisposition to fall back into the previous state under the slightest atmospheric influences. This simple inflammation of the pharynx is liable to degenerate into granular pharyngitis, especially when there is a gouty or rheumatic tendency. The strumous, gouty, and rheumatic diathesis predispose to the disease. Chronic varieties* ensue according to the constitution of the patient; thus, Sir Morell Mackenzie recognises three forms of congestion of the pharnyx (herpes of the pharynx, rheumatic and gouty sore throat), and devotes a separate chapter to each of these forms. Guéneau de Mussy, Chomel, and Pidoux attribute forty out of forty-five affections of the pharynx and larynx to herpes. Without recapitulating the previous

* Sir Morell Mackenzie, " Diseases of the Throat and Nose."

observations on the effect of our waters in these three constitutional affections, I may add that our pulverised waters, sulphurous inhalations and steam, with the internal use of Challes water, afford the surest means of attacking the malady.

CHRONIC LARYNGITIS.

(Chronic Catarrh of the Larynx.)

This affection, erroneously identified by many authors with phthisis of the larynx, consists* in a chronic inflammation of the lining membrane of the larynx, characterised by hoarseness or loss of voice, and generally by cough, more or less violent. Occasionally the malady causes thickening of the affected membranes, capable of degenerating into ulceration.

Singers, lawyers, preachers, all who over-fatigue the voice, are especially liable to this affection, as are also inveterate smokers and imbibers of alcoholic beverages. The influence of climate is likewise remarkable. In Lyons, where the situation between two rivers causes very sudden changes of temperature, these maladies are extremely prevalent.

Herpes is in this case, as in pharyngitis, the general predisposing cause. I ordinarily prescribe with excellent results general vapour baths and douches, which by their action on the skin relieve the congestion of the larynx.

For local treatment we find at Aix and at Marlioz

---

* Sir Morell Mackenzie, p. 284.

Siegle's spray-producers very well organised; the atomisation of the fluid is effected by steam. The following observations from Sir Morell Mackenzie's work have more weight than any personal views of my own :—

"Several patients whom I have sent to the Pyrenean springs have derived undoubted benefit from the use of those waters; but, on the whole, I have seen more benefit from the waters of Aix-les-Bains and Marlioz. The climate of the Pyrenees is sub-tropical, and generally very enervating in its effects on English patients. I can particularly recommend the hot sulphur waters of Savoy when the voice remains weak and the mucosa is relaxed rather than congested."

Farther on, alluding to the treatment of chronic glandular laryngitis, Sir Morell Mackenzie adds: "The sulphur waters of Aix-les-Bains are especially valuable." Apart from the testimony of this eminent London specialist, Drs. Lennox Browne, Prosser James, Smyly of Dublin, Fauvel of Paris, and others too numerous to mention, prove their confidence in the waters at Aix by sending patients thus affected, who are in most cases successfully treated.

Most valuable is the annexed opinion of Dr. Lennox Browne, who has personally inspected and studied the principal spas of Europe, and whose successive visits to Aix enable him to appreciate the value of the course.

"The only way by which pulverised liquids can be taken into the larynx and lungs, without doing more harm than good, is that in which the waters of Marlioz (Aix-en-Savoie), &c., are administered; large rooms (*salles d'inhalations*) being charged with clouds of very finely atomised medicated waters."

Later on, the same author writes:—" Within the last few years, however, great advance has been made in this branch of treatment, and it is proved, beyond doubt, that the action of natural mineral waters does not depend solely, or even to any great extent, on the amount, often very small, of active ingredients which they contain, but is the result of their natural chemical combination, and of their thermal properties. It is this last principle of natural high temperature that is to be found in almost every water of any value for bath treatment, especially of those suited to diseases of the larynx." Amongst the baths mentioned by the author, Aix-les-Bains occupies the first rank for external use, and Challes for internal use.*

Dr. Lennox Browne writes: " For the cure of a chronic cold and the eradication or diminution of the tendency to take cold, much may be done by hygienic measures, and especially by a 'cure' at Aix."†

It must be borne in mind that this course is only beneficial when the origin of the evil is suppressed; thus singers must allow their voice complete rest, and on smokers must be imposed total abstention, whilst (a far greater sacrifice in certain cases) those saloons where heated atmosphere, effluvium of gas, and, above all, moral emotions, increase the congestion of that

---

* " Diseases of the Throat." By Lennox Browne, pp. 59 and 86. Baillière, 1878.

† " Voice, Song, and Speech." By Lennox Browne and Emil Behnke. 2nd edition, p. 285. Sampson Low, Marston & Co. 1884.

delicate organ the larynx, must be completely abandoned.

CHRONIC BRONCHITIS.

Whether this affection manifests itself by abundant expectoration or in the form of emphysema, our course exercises a calming effect on the crisis. It is impossible, arbitrarily, to define when and where the warm inhalatious of Aix or those of lower temperature of Marlioz prove most beneficial.

In cases of chronic shortness of breath combined with slight cough and expectoration, the simple sulphurous vapour inhalation of Aix is generally efficacious. Patients breathe more freely on entering the rooms charged with heated vapour. Bronchitis attended by expectoration is also benefited by this warm inhalation, especially in cases of elderly people. At first, inflammation slightly increases and expectoration is more copious; but these symptoms rapidly diminish; and at the end of the course, patients, if not completely cured, find themselves far less liable to the pernicious influence of cold and atmospheric changes, and to the recurrence of the winter cough from which many chronic illnesses originate.

The internal use of tepid sulphur water at night, by exercising a gentle stimulating action on the skin, calms the dyspnœa, and soothes the cough and expectoration. Of all the remedies employed in chronic catarrh of the bronchial tubes, "steam inhalations are probably more useful than any other class of local remedies that can be employed by the patient himself. They are of the greatest service in all acute inflammatory affections of the throat, and also in most chronic affections of that organ."*

Sprays of pulverised water must never be employed where dyspnœa, in gouty, rheumatic, or herpetic bronchitis is associated with secondary symptoms. The inhalations of Marlioz prove frequently more efficacious than those of Aix on account of the mineral elements —iodine, sulphide and chloride of sodium—with which they are so highly charged. Hay fever, so prevalent among Americans, is treated successfully by vapour inhalations combined with nasal douches.

PHTHISIS.

CLINICAL as well as anatomical observations prove that this affection may be frequently cured. The tubercles are destroyed, either through elimination or by mineral transformation.† Elimination of the tubercle leaves cavities in the lung which are finally closed by cicatrisation; when communicating with the bronchi such cavities are easily recognised by the experienced ear of the practitioner. Mineralisation is a modification of the tubercle which becomes calcareous and cretaceous, remaining in the parenchyma as a foreign element, with all power of evolution suspended. M. Guéneau de Mussy believes that this cretaceous transformation occurs frequently in

* Sir Morell Mackenzie, p. 271.
† Guéneau de Mussy, tome i. p. 459

arthritic constitutions—which would explain those observations in which the tuberculous diathesis is linked with rheumatic and gouty manifestations.

During certain periods of suspension, or at least of delay, of phthisis, recourse should be had to our waters, especially when the patient is infirm or debilitated.

Better results may be anticipated when consumptive patients are of a languid, torpid, reduced constitution, than when of an inflammatory, nervous, and irritable constitution. It is important to remain within due limits, for if these are exceeded, febrile exacerbations, congestion, hæmoptysis, directly affecting the pulmonary system are excited. Not only tubercle, but all the accompanying symptoms hastening its development, are treated by our course.

Baths of short duration, taken at temperatures rather lower than that of the body, at first occasion a slight oppression and quickening of the breath, but gradually the cough is soothed, expectoration is facilitated, the pulse beats regularly, and the temperature is lowered.* After the bath the patient breathes freely, appetite increases, the pulse is lowered by twelve, twenty—sometimes twenty-eight—pulsations; the temperature falls about two per cent., perspiration diminishes or even subsides, and sleep returns. To resume, baths facilitate respiration by stimulating the action of the skin, and of the supplementary breathing functions. They also fortify the skin, render it impervious to cold, and produce on the general system

* Durand-Fardel, " Eaux Minérales de la France."

a feeling of *bien-être*, due to the bracing effect. **Douches** render great services in the therapeutics of consumption by their action in removing diathetic affections, and imparting equilibrium to the cutaneous functions. Local douches applied to the lower limbs, seconded by shampooing and friction, obviate the vicious distribution of caloric. A fluxion is produced by the increased cutaneous calorification, and improved circulation, which frequently averts local congestion, ensues.

In women, the uterus is the starting-point of congestion, and, unfortunately, in many consumptive women there is a suspension of menstruation, which may be restored by hip baths, injections, and local douches on the lower limbs. Should it then recur punctually twice or thrice, there is every prospect of a regular continuation, by which the congestive movements of the lungs are subdued.

Inhalations especially facilitate expectoration, and diminish the fatiguing cough of consumptive patients.*

The sulphur absorbed by the mucous membrane diminishes the abundant and exhausting secretions; all the affected parts thus coming in contact with the mineralised element.

Sulphur waters have often been accused of promoting hæmoptysis. Might not this complication be ascribed to the fatigue of the journey and the change of atmospheric pressure?

* The sulphuretted hydrogen, according to demonstrations of Dr. Nieper, from Alleiard, is one of the strongest toxics of the tubercular bacilli—certainly stronger than iodine, thymol, carbolic acid or sublimate.

I entirely differ from some of my colleagues in the Pyrenees, who calmly survey the advent of hæmoptysis as a natural and even beneficial result of the commencement of the course. When called on to attend such cases, after two or three days of our treatment I invariably prescribe great moderation and caution. However, when, in spite of all precautions, these symptoms continue, with perspiration and febrile indications, the course must be abandoned.

Frequently, patients unable to remain in certain spots renowned for their good effects in pulmonary affections, derive great benefit at Aix and Marlioz. The elevation of these two stations not being high, whilst the air is exceptionally pure, consumptive patients are favoured with the most important curative elements to meet their case.

SYPHILITIC DISEASES.

Our waters are renowned in the treatment of syphilis. All doctors in describing them have published observations on this class of disease. But their chemical and eliminative action on the specific poison cannot be precisely indicated, nor can any observations be relied on unless confirmed by many years of careful superintendence, as the symptoms sometimes recur at a subsequent period.

I have seen, in my own practice, a gentleman in whom, after an interval of eleven years, the old syphilitic manifestations returned after a severe course of baths and douches.

Sulphur waters do not act as a specific against syphilitic virus. Mercury alone has power to eradicate that poison.

The Aix water has an eliminative action upon the excretory system, and when combined with the Marlioz and Challes waters, given internally, has a powerful depurative action, especially in inveterate and obstinate cases.

For more than a century sulphur waters have enjoyed the reputation of revealing latent syphilis. In 1857, the Hydrological Society of Paris took this question in hand, and appointed a commission, which reported that: *Sulphur waters often reveal latent syphilis, and are useful as tests of the presence of syphilis in the system.*

Since then, in all thermal stations this law has been confirmed by careful observations. Many medical men of Aix (Savoy), Carlsbad, Aix-la-Chapelle, Bagnères de Luchon, Schinznach, &c., have published interesting illustrations of this fact.

Dr. Hermann, of Schinznach, terms the sulphurous water the reactive against syphilis, and although not infallible, this view is to a certain extent confirmed by my experience. While admitting the legitimate use of mineral sulphurous water in this important point of diagnosis, I believe that rare exceptions occur, especially when a very long treatment has not been undergone.

The duration of the cure cannot be limited to the fashionable course of one-and-twenty days. A long, graduated, and interrupted treatment is essential, in order to obtain the double result of excitation and manifestation. The latent state exists, especially in lymphatic and strumous patients deficient in reactive energy.

Nowhere better than at Aix can be found the means of graduating the temperature and sulphurisation of the douches and vapour baths, so as to stimulate cachectic and feeble constitutions.

We must pay special attention to the thermality, which is the sole bond of union between the water containing chloride of sodium used in Germany against syphilis and the sulphurous springs employed with the same beneficial results in France and Spain. The Spanish hydrologists assert that the springs of Murcia, Carratraca, &c., are in themselves specific against syphilis.

After the internal use of Challes water, so rich in mineral elements, secondary symptoms frequently disappear.

The treatment at Aix acts energetically against mercurialism, which often complicates the syphilitic cachexy. The water retards the salivation produced by mercury and all other mercurial phenomena.

The elimination of mercury by sulphur has been long recognised. Dr. Blanc published many years ago a pamphlet on the elimination of mercury by the skin resulting from the use of the sulphurous medication.

I have often observed the presence of mercury in a metallic state in the urine of patients after some days of treatment.

Hermann had already detected mercury in the stools of his patients affected by mercurialism.

Durand-Fardel offers the following explanation of the action produced by sulphurous water:—"It can be obtained from nearly every mineral water at high temperature, whether the mineralisation be weak or strong, but it especially belongs to thermal sulphurous water."

If some specific symptom appears, we prescribe a severe internal or external specific medication, to be followed at intervals during some years, as urged by Professor Fournier, until the patient is convinced that a permanent cure is effected.

SCROFULA.

Cases of scrofula are most frequently met with among the young people resorting to our baths. The sulphurous waters of Aix, seconded by the internal use of Challes water, exercise a stimulating, invigorating, and restorative influence. During many years the beneficial action of the Challes springs was ascribed to the slight quantity of iodine contained in them; further researches have demonstrated that it is due to the presence of sulphide of sodium, which has the double effect of increasing the power of nutrition and diminishing the glandular exudations, thus permitting a better circulation of lymph.

We do not believe in the alterative anti-scrofulous properties attributed to certain German baths. We know the effect of chlorine on blood as a chemical experiment; but the general action of mineral water on the entire system is not that of a chemical agent. The *modus operandi* generally depends on the association of different chemical principles, and on the application of the hydro-mineral system of treatment.

The vitality of the skin is thus increased, the nerves and muscles are strengthened, all the functions are stimulated, especially those of the lymphatic system. We succeed chiefly in the torpid forms of scrofula.

Amongst the numerous cases of scrofula cured by our waters, I will only mention two cases of erythematous lupus of the face, and three cases of lupus rodens, completely cured after two months of baths and spray of Challes water.

The greatest proportion of scrofulous affections at Aix are manifested in disease of the bones and articulations; in the latter case the injury probably commences in the synovial membranes or in the epiphyses. The surrounding parts are tumefied and fungous, and every sort of affection, from simple abscess to groups of fistulæ, from stiffness of articulation to subluxation, may be encountered. Femoral affection is unfortunately the most frequent.

We will not repeat our general observations on articular diseases, but only add that constitutional scrofula aggravates them to an alarming extent. Hence arise *caries, necrosis,* or simply *osteitis* and *periostitis,* with abscesses and fistulæ in the articulations. Douches, and especially vapour baths, stimulate the granulation of the ulcers, of the bony and fibro-muscular tissues, distending and eliminating the degenerated pieces of bone. Frequently, this elimination occurs long after the cure which produced it.

M. Petrequin has demonstrated the efficacy of our course in cases of scrofulous ophthalmia.

The alterative and tonic action may be manifested, but not completed, in the short space of twenty-one days in deeply rooted affections, such as scrofula. Long and careful treatment under medical guidance can alone eradicate the disease.

SKIN DISEASES.

THE local effect of the bath is a simple *pansement* of the dermatosis ; but it also acts generally on the nervous system, and if the cutaneous affection arises from constitutional disturbance, such modifications may ensue as to ensure thorough recovery. The greater the absorption of the sulphurous element, the more hopeful are the prospects of a cure. This absorption, slight when the skin is dry and scaly, is very great in moist affections where the mucous parts are exposed to the air.

Skin diseases may be of rheumatic or gouty nature. Sulphur waters only act in chronic cases ; of those resulting from rheumatism, such as eczema, herpetic erythema, acne, lichen ; and those due to gout, such as psoriasis, eczema, prurigo, acne, &c.

The inveterate form of **psoriasis** is the commonest at Aix, and results equally from gout and rheumatism. Our waters unaided do not effect the cure ; but under their influence the patches become fainter, the scales fall from the skin, and the absorption and action of internal medicines are facilitated.

I never continue the course in those cases where acute symptoms reappear, but I remember three recent cases of torpid psoriasis of rheumatic origin which were completely cured at Aix.

Prurigo results from want of vitality in the skin. The mineralisation of our baths suffices to arouse the torpid functions without irritating the skin: I have treated successfully, by means of simple tepid baths, patients covered with pimples which caused unbearable itching.

In the article to which I have referred, Sir A. Garrod says:—"The skin affections which I have for the most part seen treated at Aix have been those in connection with rheumatoid arthritis and gout. I have usually seen psoriasis associated with rheumatoid arthritis, and eczema with true gout. In regard to the occurrence of psoriasis in rheumatoid arthritis, I am sure that the Aix treatment, in relieving the articular, usually ameliorates and often cures the skin affection. I cannot help quoting here a case illustrating this statement which came under my care some years ago. An elderly man, a high dignitary of the Church, had for many years suffered from psoriasis over much of the body, and on the back was a patch larger than any I had before seen, covering the whole surface of that region; he had more recently developed joint symptoms, evidently of rheumatoid arthritis. Although somewhat old for mineral water treatment, still he was suffering so much that I sent him to Aix, chiefly for the purpose of arresting the somewhat rapid increase of the articular disease. After about a month's treatment he had improved in general health, but had not made much progress as far as the skin was concerned; but during the next eight months after leaving Aix the skin disease rapidly improved, and the back became quite free, and, in fact, the amelioration effected both in joint and skin appeared quite marvellous. I have observed in numerous instances that the eczema occurring in gouty patients becomes much relieved by an Aix course, but in some few cases I have been disappointed."

Eczema.—Sulphur has long been considered the specific cure of the malady, but I am inclined to modify this view, and not to employ sulphur waters (Aix, Challes, or Marlioz) in excess, when their action stops too rapidly the progress of eczema. I once saw a child nearly killed by the application of Challes water for three days, which suddenly stopped an attack of impetigo capitis of the scalp. The first effect of sulphur baths on eczema is to exaggerate the symptoms; but if the treatment is persevered with, the manifestations gradually disappear; our hospital patients derive the greatest benefit in these cases, because they do not require to hurry the course, and their tonic treatment is thus unaccompanied by fatigue.

Acne.—Here the tonic and stimulating swimming baths are most efficacious, with local spray directed on the part affected. The suppuration and cicatrisation of the small pustules are hastened without leaving hard red tumours.

Thus it is evident that the category of dermatoses influenced by our waters is limited, but in those referred to their local effects and tonic action are very remarkable.

DISEASES OF WOMEN.

Simple and Follicular Vaginitis.—The irritation so troublesome in this disease, when connected with eczema, or due to a rheumatic tendency, is greatly alleviated by the use of sulphur baths. Frequently the sebaceous follicles of the mucous membrane of the vagina are inflamed, and produce disturbances of the general health, with discharges, pains in the back and thighs, constriction of the sphincter, and great discomfort in walking, the uterus remaining in its normal state. I have invariably treated these cases by a course of plunge baths of Aix water, of hip baths of Marlioz water, and injections of Challes water, especially avoiding cold vaginal injections. The mode of administering injections requires skilful attention. Two years ago I performed on one of my patients affected by vaginal leucorrhœa the excision of a series of granulations so hard and voluminous that I feared for one moment they were of the nature of epithelial cancer. Afterwards I was enabled successfully to introduce a cannula and a bath speculum. This treatment lasted forty days. I saw her six months afterwards; the leucorrhœa and mental depression, previously present, had completely disappeared.

Hæmatocele.—A remarkable instance of a lady afflicted with hæmatocele (retro-uterine) was sent to me by one of the leading physicians of Paris.

Mrs. M———, aged forty-four, periods regular, feeble constitution and circulation; one child; complains, after walking, of violent pain in abdomen; is sometimes sick, frequently requires to empty the bladder, had itching of the rectum, small tumour in the pelvis appreciable through the vagina in the posterior wall. This was her state when first I saw her on the 2nd of August 1880. I immediately prescribed a course of simple sulphurous baths at 96° to be taken every morning, with hip baths and slight vaginal injections of Challes water during twenty minutes at night, and two tumblers of Challes water to be drunk daily. After thirty-four baths Mrs. M——— left Aix completely cured. I often see her in Paris; the periods are regular, the micturition is quite normal, and she has not experienced any return of the symptoms. This cure was due to the alterative properties of Challes water, and to the general stimulating effects of sulphur baths.

Uterine Diseases.—We will only allude to *Amenorrhœa* and *Dysmenorrhœa*. When these diseases are not produced by natural deformity requiring surgical intervention, they result from general conditions to which the stimulating action of our waters affords certain relief. To women of a certain age I prescribe douches and shampooing. Young girls should resort to the swimming bath, as that invigorating exercise adds to the stimulating effect of the waters.

Metritis—Ulceration — Leucorrhœa.—The study of uterine affections proves that the greatest number result from chronic metritis, which, except in a few traumatic cases, starts from some kind of diathesis. If, according to eminent pathologists, the greatest proportion of cases of laryngo-pharyngitis may be traced to dermatosis, that association is also frequently met with in affections of the uterine mucous membrane, such as follicular catarrhal metritis and consecutive leucorrhœa. Displacements and all their neuro-pathogenic concomitants, such as anæmia, nervousness, dyspepsia, insomnia, weakness, and weight in the abdomen, are under the influence of uterine disorder. It is most interesting to follow the evolutions of this diathesic state, sometimes localised in the vaginal or uterine mucous membrane, sometimes producing bronchitis or acute angina, which in turn give place to eruptions of the capillary or other portions of the cutaneous covering. Another cause of congestion of the uterus, and, hence, of all disorders resulting from the lesion of so delicate an organ, is the rheumatic diathesis.

The baths and douches of Marlioz are adapted to the purely herpetic state; those of Aix to rheumatic complications. Our medication, whilst determining a revulsion on the entire cutaneous surface, re-establishes the functions of the skin. " By intense sudations it reduces the morbid tendency, and substitutes perspiration by the pores for the leucorrhœal discharge." *

Experience has proved the pernicious influence of cold water on these delicate organs. I remember, amongst others, the case of a lady suffering from chronic uterine catarrh, who, after taking three cold hip baths, and three cold local injections, was attacked by a very serious metro-ovaritis.

The conclusive arguments of Virchow† demonstrate that the contraction of the arterial system due to cold leads to congestion of the internal organs, and especially to that of a uterus if already enlarged.

We are by no means surprised when consulted by women who return from sea baths and hydropathic stations, with an exaggeration of those symptoms which accompany metritis—such as a sensation of heaviness and weight in the hypogastric, uterine colics, dysmenorrhœa, throbbing with tenderness about the groins and perineum, hyperæsthesia, and itching of the vulva.

The abuse of strong and repeated injections is also pernicious; since there are cases in which injections increase the natural irritability. Therefore we substitute the use of specula of gutta percha or of metal in the bath, as they can be easily introduced and kept in place by the patient; and the organ thus profits by the bath without the irritation of a

---

* Courty, "Maladies Uterines."
† Virchow, "Physiologische Beschwerungen über das Seebaden," page 89.

sudden shock, or of a different temperature produced by the injection.

We have especially noted the advantage of bathing of the womb in ordinary cases of tumefaction of the glandulæ and follicles of the cervix. The patient feels how far she can introduce the instrument without tearing the sensitive granulations. " The thermal treatment of chronic metritis, consisting of prolonged baths and swimming baths, is particularly efficacious. Vaginal irrigations modify catarrhal surfaces, but vaginal douches should be almost entirely abolished, as they excite congestion and uterine and peri-uterine neuralgia."*

M. Denos, physician of the Hôpital la Pitié,† advises the powerful application of therapeutic agencies in cases of torpid metritis, and metritis complicated by the rheumatic or gouty diathesis.

We have frequently observed, as an equally important result of a course of baths, the power of stimulating the effects of local treatment previously abortive, especially in cases of torpid metritis complicated by scrofula. When the reaction is thus effected, astringent, emollient, or caustic remedies may be applied according to the case.

Displacements and prolapsus, generally produced by metritis, disappear when their cause is removed. Tonic sulphurous treatment acts also most efficaciously on relaxation of the ligaments.

First Case of Ulceration of Cervix Uteri cured at

---

* Durand-Fardel, " Leçons à l'Ecole Pratique," p. 212. 1874.
† "Annales de Gynécologie." 1876.

Aix.—Madame C——, twenty-five years old, of good constitution but of lymphatic temperament, arrived at Aix, June, 1884. She commenced menstruation at the age of fifteen, without pain; married at eighteen, she had a child fifteen months afterwards, and subsequently suffered acute agony in the hypogastric regions, with white discharge after the monthly epochs; pelvic and sacral pain, increasing in intensity whilst walking and driving; and great mental depression. There was an evident congestion of the cervix; and I observed through the speculum an indolent bleeding ulcer of no great depth. Madame C—— had been cauterised eight times without result; as there appeared to be no displacement of the organ, I recommended simple swimming baths. After the fifth, the patient complained of a discharge of mucopus, and consented to my proposal of a general cauterisation of the ulcerated parts. Three days afterwards I prescribed one tepid hip bath and injection, to be followed by the ordinary large baths, with an internal bath by means of the speculum.

By degrees Madame C—— was enabled to walk in comfort, the monthly periods occurred in their usual course without suffering; and after thirty baths she was completely cured. I advised her to take walking exercise and frequent tonics, and invariably to rest after experiencing the slightest fatigue. She has been very well ever since.

Second Case: Chronic Metritis proceeding from Rheumatism.—Madame V——, twenty-three years old, married two years, of weak constitution and lymphatic temperament; had inhabited ever since her marriage a damp new house; had no family. Three months after her marriage, whilst suffering from dysmenorrhœa, she was seized with acute pelvic pains and violent backache, which she attributed to a drive in an open carriage on a damp cold day. Treated by a midwife, whose diagnosis she could not produce, she had taken baths and injections, with applications of blisters and leeches to the hypogastric region, and thus passed eighteen months without experiencing the slightest relief. She arrived at Aix in the beginning of the summer of 1874 with her general health greatly impaired and with difficult digestion; the chronic pains increased whilst walking or sitting upright; there was irritability of the bladder, and fulness and heat about the pelvis; the cervix, which was pushed forward and prolapsed, was hard and sensitive. The uterine catheter entered easily to a distance of about six centimetres. In this obvious case of uterine and peri-uterine inflammation, I prescribed iodide of iron before meals, a tumbler of Marlioz water at night, and daily tepid Aix baths with the speculum bath, the case being of too inflammatory a nature to support more stimulating injections. After seven baths, as the patient complained of pains in the shoulder and wrist, I deemed it prudent to discontinue the treatment, and the pains ceased. I then recommended hip baths, accompanied by douches to the back and shampooing of the limbs; the patient then to be swathed until gentle perspiration ensued. After eight douches the monthly periods returned in greater abundance, and of more satisfactory colour, necessitating five days of complete repose. An examination showed great local improvement; the cervix, previously hard and swollen, had become supple. By my directions the douches were resumed and mild irrigations of sulphurous water were substituted for the speculum bath; this treatment, with occasional interludes of tepid baths, was continued until the next period, which passed without the slightest symptom of pain. After two months, during which Madame V—— had taken twelve baths and eighteen douches, she left Aix, her local as well as general state completely re-established.

Disturbances at the Epoch of Change of Life.—The chapter of women's diseases would be incomplete without a reference to those illnesses occurring at the critical period which are advantageously treated at Aix. The functions of the skin are in sympathy with all organic, and more especially the uterine, functions. Therefore, when the latter are suppressed, the entire cutaneous system is affected, according to the special tendency of each individual, to congestion of the brain, neuralgia, sick headaches, dermatosis, rheumatism, gout, hæmorrhoids, congestion of the respiratory organs, &c. Our general bathing system, our douches and shampooing, by improving the circulation of the blood through the capillary surface, avert these critical and often dangerous symptoms, whilst they stimulate the general organic functions.

HYSTERIA.

A VARIETY of pathological conditions, vaguely denominated nervous, to which women are liable, are connected with hysteria. They cannot be defined, because, in spite of their existence and the havoc they produce, no specific effect is observable on the organic system. The following are the usual symptoms among women frequenting our baths: impatience, anxiety, insomnia, vague floating pains, feelings of suffocation, convulsive outbreaks of crying or laughter, occasionally with hiccough; temporary weakness of a limb; tacit resistance to exercise; often loss of appetite, accompanied by dyspepsia; more rarely digestive disturbances leading to rapid decline. It is very difficult to define the causes of these affections during the course of twenty days' medical attendance; neurotic affections may be engendered by the rheumatic, syphilitic, or herpetic diathesis, by functional troubles or uterine affections, or by mental anxiety; more frequently they are connected with a cutaneous hyperæsthesia resulting from torpid functions of the skin. We have already referred to the satisfactory effects of our treatment in diatheses, as well as in anæmia and uterine diseases. Locally, ascending cold baths, with general douches of short duration and slightly higher but still tonic temperature, often produce the best results. When neuropathy is allied to a disturbance of the cutaneous functions, how can a more efficacious mode of judging the case be found than our Scotch douche?

Referring to the work of Dr. Playfair and the system

prescribed by him for hysteria, which I published in French, I may mention that at Aix are found thoroughly trained shampooers, and that women who resort to Aix for treatment, can be completely isolated from their families and general surroundings, according to the conditions imposed by Weir Mitchell and Playfair.

ANÆMIA AND CHLOROSIS.

By some anæmia has been considered as a symptom, by others as a specific illness; but all agree that it results from poverty of blood and excess of its watery element. Anæmia evidently comprises chlorosis, which is one of the forms in which it attacks young people. As anæmia is a morbid condition of the blood, all the tissues and organs are thereby affected. In anæmia, of whatever origin, our douches—of which it is easy to modify the strength, temperature, and sulphurous element—produce the best effects.

Children and young girls, who arrive at Aix, pale, haggard, bloodless, assume in a very few days a healthy aspect. The doctor directs the thermal course according to the impressionability of the patient, avoiding both low temperature, as reaction is difficult in bloodless natures, and very high temperature, which tends to increase the general enervation. Where atony precludes reaction simple baths of short duration should be taken; then swimming baths; and douches, with gentle shampooing, as soon as the exciting properties of our waters have established reaction. Sweating and perspiration should be avoided, not on account of any risk from diaphoresis in anæmic cases, but in order to induce the patient to bring about a reaction for herself, thus, by a physiological effort, producing a more effectual tonic action upon the skin. Envious critics have vainly sought to injure the reputation of Aix by attributing abundant diaphoresis to the hydro-thermal treatment. "For profuse perspiration cannot provoke adulteration of the blood; physiology forbids us to believe in anæmia caused by perspiration, and clinical observations confirm these negative data." *

These waters have no actual anti-anæmic or anti-chlorotic qualities, such as belong exclusively to ferruginous waters. But how rarely can anæmic patients support the tonic treatment of iron, which is constantly impeded by gastralgia or by obstinate constipation. External sulphurous treatment is easily supported, and I always prescribe the waters of La Bauche, Orezza, or Bussang, to be drunk when they can be assimilated. Many of my patients, unable to endure the stimulating effects of sea baths, or the congestive effects of the spas of Spa, Schwalbach, or Franzenbad, derive benefit from our *Douches en cercle* and swimming baths accompanied by sprinkling with cold water.

---

* Professor Lee, "Leçons de Pathologie experimentale," p. 86. 1867.

MYXŒDEMA (of Ord)—CACHEXIE PACHYDERMIQUE (of Charcot).

This malady, which was first designated by Sir William Gull as a cretinoid state affecting women in the adult period of life, was subsequently called myxœdema by Dr. Ord, and *Cachexie Pachydermique* by Prof. Charcot. Its etiology has been of late the subject of serious study in France and England; and in 1882 was exhaustively discussed in several meetings of the Clinical Society of London, by those eminent practitioners—Heron, Goodhart, Marcet, F. Taylor, Hadden, Seymour, Ord, Taylor, Dyce Duckworth, Mahomed, Cavafy, and Haward.* The publications of Hadden,† and the interesting notes of MM. Reverdin, Professors and Surgeons of Geneva,‡ throw fresh light on the origin of the disease, without satisfactorily solving the problem. Dr. Mayor of Geneva, after analysing all previous publications, concludes that it is a lesion of the sympathetic nerve.§ It would be superfluous to detail the symptoms of an illness now so generally understood, but the observations of MM. J. L. and Aug. Reverdin and of Kocher of Berne on the phenomena identical with myxœdema, met with in cases of goître, in which the thyroid gland had been entirely removed, are too novel and interesting to be entirely passed over. Dr.

* *British Medical Journal*, 1882, pp. 80–424.
† Hadden, "The Nervous Symptoms of Myxœdema."
‡ "Vingt et deux Opérations de Goîtres," *Revue Médicale de a Suisse*. Romande, Juin, 1883.
§ *Revue Médicale de la Suisse*. Romande, Septembre 15, 1883.

Ord had already called attention to the alteration of this gland, which, according to him, is usually atrophied; and it is curious to follow the connection discovered by MM. Reverdin and Kocher between the thyroid gland and the malady. Certain observations on myxœdema tend to prove the atrophy and alteration of the thyroid, whilst, on the other hand, the surgeons who have removed that gland affirm that three or four months after the operation symptoms of myxœdema have manifested themselves in their patients. "Something analogous must occur in the sympathetic nerve to what is produced in the central nervous system, or spinal cord, whereby muscular atrophy—primary or secondary—induces peripheral lesions."*

Some doctors have called attention to the marked diminution of urea in the urine of these patients. Sir Andrew Clark wrote to me two years ago, when I was preparing my pamphlet, "that as Dr. Ord was the first to publish his observations, he did not proceed further with some notes collected and prepared by Dr. Burnet and himself, establishing that the fundamental factors of the disease are imperfect kidney, combined with a failing capillary circulation and feeble heart, and that it does not take its origin in the nervous system, which only afterwards becomes affected."

Whatever may be the origin, it is a cachexia without abnormal proliferation; an intellectual and physical depression, in which, since the skin is first and

* *Revue Médicale Suisse*, p. 522. Mayor.

most seriously affected, an exciting and stimulating external treatment becomes imperative. Three cases in which, after a long period of attendance, I fortunately obtained satisfactory results, may prove interesting, as I was the first to employ our course in this affection; and presented my observations on their salutary effect to the Scientific Congress of Aix in September, 1882.

First Case.—Madame—— V was sent to me in 1878, by Prof. Charcot, with this diagnosis: (*Cachexie pachydermique*); aged forty years; born in the Island of Zante. This woman presented a face in the form of a full moon, with a vacant expression, purple cheekbones, skin of a waxen hue, eyelids swollen and hard to the touch, the nose wide at the root, snuffling voice, hands and feet swollen, the hair fallen from the armpits and pubes, head almost bald, slow and measured gait. There was intellectual indolence and heaviness, difficulty in expressing herself, almost complete forgetfulness of English and Italian, which she had spoken perfectly. The tongue was thick, and of a violet hue, and the tongue muscles were very indolent. The periods were regular. The stomach acted badly and in a *bizarre* manner, only white wine, slightly acid, being tolerated. From these combined causes a profound anæmia had resulted. The patient was, moreover, melancholy and very irritable.

Immediately attacking the defective functions of the skin, I submitted Madame V—— to a series of sulphur baths, alternated with vapour baths. From the beginning the scales fell from the skin, and did not reappear. This first success encouraged the patient; and she subsequently took two courses of thirty days each.

In 1880 Dr. Thaon, of Nice, thus commented on the case in the *Revue Medico-Chirurgicale* :—" The patient is much better after passing two seasons at Aix. Most of the symptoms are improved ; the face, eyelids, and lips are less swollen, the skin of the extremities is less hard, speech less slow, and the patient is enabled to take long walks."

Three years have elapsed since these lines were penned, and Madame V—— passes regularly three or four months at Aix, continuing the vapour baths, alternated with sulphur baths. When last she left—a few days ago—the œdematous condition of the mucous membrane and the skin had disappeared, speech was far more articulate, and the organs of intelligence and memory had participated in the general improvement ; in proof of this she again speaks the languages learned in her youth. This cure is entirely due to the external application of Aix waters, for this patient positively refused every kind of medicament.

Second Case.—Madame N——, sent last July by Dr. Ord : aged thirty-four ; had been a very healthy child (no hereditary or personal antecedents known) ; regular menstruation from the age of fifteen. The illness first manifested itself six years ago by spongy sensations in the fingers, difficulty in writing, pains in the side, flatulence, and general debility ; then gradually the hair fell from the body, the skin became scaly, the eyes swollen, the tongue thickened, speech heavy and embarrassed, whilst intelligence diminished greatly. In this state she first arrived. I submitted her to a similar course to the one quoted in the preceding observations—alternate sulphur baths at 96°, and vapour baths. After a month the skin had resumed its normal state, but the mucous membranes were in no better condition, nor was speech less difficult; muscular and intellectual activity were completely restored.

Third Case.—Madame J——, forty-four years of age, was sent to me in August 1882, by Professor Charcot. Born and living in Canada, she had had one daughter, in perfect health, by whom she was accompanied. Madame J—— had never been ill until about two years previous to her arrival. Her eyelids had suddenly become swollen to such an extent as almost to conceal her large and well-shaped eyes; the face assumed the aspect of a waxen mask in the shape of a full moon (Gull). The cheekbones were not deeper in colour; the mouth was tumefied, the tongue enormous and of violet colour, the voice nasal and thick of utterance; it was impossible to take between the fingers any portion of the tissues of the face, which appeared stretched as if covered with a layer of collodion. Intelligence troubled and depressed; temper impatient and irritable; she scarcely took cognisance of the conversation addressed to her. Teeth in dilapidated condition; feet and hands massive; constipation obstinate. At first I encountered steady disinclination to follow my prescriptions, which rendered the treatment difficult. By dint of gentle persuasion I ultimately induced her to take baths, vapour douches, and spray on the face; and to this treatment the remarkable appearance of the waxen mask yielded, the swelling in the eyelids diminished, the eyes resumed their normal form, and the skin of the face the usual elasticity. After forty days the patient departed, with full confidence in the successful issue of the treatment.

PARALYSIS.

The total or partial loss of sensibility or motion, or of both, in one or more parts of the body, constitutes a disease adapted to the treatment of Aix, especially when it is due to alteration of the blood, to eruptive, typhoid, intermittent or continued fever, to diphtheria, reflex paralysis, &c.

Of this I will give but one instance:—Madame B———; aged twenty-seven; of good constitution, and the mother of three children; was attacked by severe diphtheria, which produced a complete paralysis of the limbs eleven months before her arrival at Aix in July last. After two months' course of douches and shampooing, attended with applications of continuous currents, she left completely cured.

Local Paralysis.—Although not admitting complete rheumatic paraplegia, of which I have not met with a single instance, I observe in rheumatic and gouty people, after exposure to cold or to sudden changes of temperature, some local incomplete palsy, such as the facial paralysis of Bell, palsy of the muscles of the lower extremities or of the shoulder, which can only be accounted for by these external influences. Many of these cases, which yield readily to our treatment, occur in my practice.

Hysterical Paralysis.—When I observe palsy of one lower extremity in hysterical persons, I resort to tepid baths, followed by gentle shampooing whilst the patient is in bed.

Mercurial and Lead Palsy.—Having already alluded to the excretory power of our water, due both to its chemical and physiological action, it follows that in lead palsy our treatment is eminently successful. In my practice I have met with two cases of complete recovery from palsy of the muscles (extensors) of the hands and fingers resulting from lead poisoning.

In a case of palsy of the fifth nerve due to mercurial poison, I was equally successful; after a few days mastication became quite easy. Combined with the specific eliminating power of sulphurous water, the shock of the douches, and the effect produced on the muscles and capillary circulation by massage, constitute an almost infallible cure.

When paralysis or palsy is due to organic alteration of the brain, softening, induration, tubercular or cancerous, or to renal degeneration, improvement may be obtained, but never a complete cure.

SYPHILITIC PARALYSIS.

PARALYSIS or palsy may be due to syphilitic gummata. Without recurring to the treatment of the specific diseases, I only observe that the course of vapour baths or douches promotes the efficacy of medication, and facilitates the resolution of the gummata.

HEMIPLEGIA.

HEMIPLEGIA, which is often due to cerebral hæmorrhage, having for its seat the corpus striatum and optic thalamus, affects the side opposite to that on which the lesion in the brain occurs; the face and limbs may alike be affected.

Hemiplegia produced by apoplexy is sudden, and may be greatly benefited, if not entirely cured, by this treatment; but the fitting moment is determined by the nature of the inflammatory symptoms and their effect on the constitution of the patient. The sooner it can be resorted to the more rapid is the resolution of the clot. When after two or three months congestive phenomena have disappeared, the cure obtainable by these different modes is commenced.

1. By re-absorption of the watery part of the blood, and solution more or less active of the fibroma.

2. As the organ becomes habituated to the compression of the clot, the nervous circulation gradually returns and recovery ensues, notwithstanding the existence of the clot.

3. Such fibres as have been divided, lacerated, or compressed by the blood, may be replaced in their functions by the adjacent fibres, just as one small artery replaces another where a ligature has been applied.

The dissolving and stimulating action has a share in each of these modes of recovery. During the last twenty years many satisfactory results have come under my notice, of which I give one very interesting case.

Hemiplegia of five months' duration—Complete Recovery.—Captain P——, aged fifty-five, of strong constitution, without any previous disease, arrived on the 20th of June 1874. During February he had a sudden attack of apoplexy, followed by hemiplegia on the right side; arm more completely paralysed than leg; sensation almost lost; tongue implicated; the joints turning to affected side; tendency to shed tears. The patient was submitted to slight douches and shampooing, which he bore very well, and left after twenty-five days much improved. He returned in August and recommenced the same course, after which he started for home, completely cured, able to write and shave.

In these cases I prefer the douche called *des princes neuf*, where the vapour is not accumulated in the douche.

PARAPLEGIA.

THE same must be said of paralysis of the lower limbs. First ascertain whether it arises from disease of spinal cord or membranes—spinal meningitis, myelitis, congestion, softening, tumour, syphilitic disease—or whether it is reflex—that is, due to irritation reaching the cord from a sensitive nerve. When blood is deficient in the cord (Brown-Séquard), diagnosis is necessary for prescribing a proper course of baths or douches—no easy task. The treatment succeeds in reflex paraplegia when it is able to attack the origin of the malady; and thrice in my practice good results have occurred in cases traced to uterine disease. In reflex paraplegia the stimulating and invigorating course is always prescribed.

Paraplegia due to Myelitis, Meningitis, &c.—The treatment of Aix is strongly indicated in these cases. The nutrition of the limbs is maintained by shampooing, stimulating douches or vapour; and gentle sprinkling of tepid water on the spinal regions.

Diseases of the nervous system constitute two-fifths of the illnesses seeking relief at our Baths. "Le traitement des Paralysies est aussi un des fleurons de la couronne thérapeutique d'Aix." *

In conclusion:

1. Our medication, however frequently used, never produces exacerbation.

2. All forms of paralysis may be treated at Aix immediately after the disappearance of inflammatory symptoms.

3. It never produces injury in chronic paralysis.

4. It stimulates nervous action and absorption of the clot.

5. It is especially successful in specific paralysis.

6. Recovery is most certain in cases of palsy due to injury or to special rheumatic or gouty conditions.

Often, in combination with the thermal course, I employ galvanic currents.

* Lombard (of Geneva).

SURGICAL CASES.

I PRESENTED to our College of Surgeons in Paris a series of observations based on certain cases of the wounded during the late Franco-German War, tending to prove that our waters applied as indicated are a sovereign remedy in such injuries.

FRACTURES.

THE bony callus is strengthened by our treatment, but only after the last period of its formation. We recommend in these cases local vapour baths, followed by a slight massage of the muscular part, as the stimulating action is first manifested on the external tissues, sometimes to the injury of the deeper ones. The results of fractures are treated with advantage at Aix—atrophy, retractions, pains, ankylosis, swelling benefiting most by this course.

DISEASES OF THE JOINTS AND BONES.

Joint Stiffness — Periostitis — Fistula — Ulcerations — Necrosis.—Referring to the classification of William Adams of those cases in which forcible movements are useful, I particularly recommend a previous course of our vapour baths with shampooing:

1. Cases of traumatic origin in healthy constitutions, generally occurring in the adult.
2. Cases after rheumatic inflammation of the joint.
3. Cases after strumous disease of the joint.
4. Cases after acute suppurative inflammation of pyæmic origin; and suppurative inflammation in the neighbourhood of, and extending to, the joint.
5. Cases consequent upon muscular contractions.*

Our local steam-baths are most efficacious after those operations (forcible movements) successfully performed by Drs. W. Adams, and Wharton Hood.

Caries proceeding from scrofula, from other general conditions, or from local influences, is much improved under our treatment. The baths eradicate the constitutional disorder; the local vapour douches with slight massage improve the nutrition of the diseased part whilst encouraging the capillary circulation and the tissue changes. The organic matters (glairine and barégine) contained in our waters have a more beneficial effect in these bone diseases than any other external applications. The elimination of the pieces of bone is promoted, and cicatrisation of the tissue ensues. The course must be very carefully watched, and all precautions used to avoid the risk of producing inflammation of the affected parts.

---

* "Selection of Cases for Forcible Movement in the Treatment Stiff Joints." By Wm. Adams. (Lecture, August 1882.)

ILLNESSES FOR WHICH THE WATERS OF AIX ARE UNADVISABLE.

TWENTY years of local experience have confirmed the views of our predecessors of the past century in respect to those maladies which ought not to be sent to Aix for treatment, for our waters do not exercise universal cures. Great as is their efficacy in the maladies enumerated, their stimulating effect is equally prejudicial when tending to excite and increase the morbid elements.

1. We have seen that the waters of Aix are most effective in treating **torpid phthisis,** but they must be avoided in congestive phthisis complicated by intense febrile symptoms and repeated hæmoptysis. In these cases I have never witnessed the amelioration traced by certain doctors in bathing-places of a similar nature to ours.

2. **Cancerous Diseases.**—Neither Aix nor any other mineral water produces beneficial action on the heteromorphic tissues. I have even observed in three cases an alarmingly rapid increase of cancerous formations in patients affected by cancerous cachexia after taking a few douches or simple baths. Once I observed in a woman at a critical period an invasion of cancer in the breast; in the second case it was an epithelial cancer of the œsophagus; a third time it was a cancerous tumour of the stomach; the progress was always most rapid; all the virulence of the cancerous cachexia was aroused and concentrated on the

spots affected. During the last twenty years a reaction has set in against the opinion that the waters are injurious in all cases of heart affection. Personally, I find in my daily practice, as well as in the service of the hospital, cases of rheumatism, with cardiac complications, which yield to the influence of the treatment. Some years ago I called the attention of some of my colleagues to a young child, aged eleven years, in whom the most inexperienced ear could detect severe pericarditis. The motion of the heart was tumultuous and perceptible from a distance, the respiration was hurried. I never heard so loud a systolic bellows murmur, nor so well marked a friction sound. Commencing with mild tepid baths, I proceeded to administer slight tepid douches with shampooing. After a month of this gentle treatment, with frequent intervals of rest, she left completely cured, and with hardly any increased intensity of the natural sounds.

I am happy to find myself in agreement with my colleague, Dr. Blanc, who holds that our course is not contra-indicated in cases of heart affection, such as pericarditis and endocarditis resulting from rheumatic fever. Evidently when there is a tendency to syncope or œdema of the face or extremities, the treatment is dangerous, as also in atrophy or hypertrophy of the heart; in dilatation, or aneurism of the heart and of the large vessels, and in angina pectoris. The best mode of diagnosis in these cases is by Dudgeon's sphygmometer, which is portable and easy to apply to the radial artery; or by

Dr. Oliver's cardiograph, which is equally portable and easy of application. Again, when nervous irritability or congestion of the brain is present, our treatment is contra-indicated.

SPRINGS OF MARLIOZ.

MARLIOZ has justly participated in the development of Aix, since every effort has been made to improve and utilise its resources. The springs, more sulphurous than any met with in the Pyrenees, yield an average of twenty thousand litres daily. Iodine and the chlorides are invaluable in diathetic cases, while manganese is an excellent reconstituent.

The water of Marlioz has a great effect on the urine, which it renders alkaline, and purifies from the uric acid produced by certain affections of the bladder; this alkalization extends also to the perspiration and other secretions. The general action resembles closely that of the waters of Aix.* But the difference of temperature and of the proportion of free sulphuretted hydrogen must be taken into account.

At Marlioz, as well as at Enghien, Eaux Bonnes, Allevard, &c., are found the best therapeutic apparatus for respiratory and uterine diseases, and the most comfortable installation for baths and douches of mineral water of every degree, with natural cold water to meet all hydropathic indications.

The waters of Marlioz act in the same way as those

---

* See the "Analysis of the Waters of Aix."

of Pierrefond, Enghien, and Allevard in affections of the respiratory organs.

Having indicated their general properties, I only add a careful analysis made by Bonjean of the waters some years ago.

TEMPERATURE, 57° FAHR.

In 1000 grains of water:—

Sulphuretted hydrogen	6·70 gr.
Carbonic acid	4·64 ,,
Iodine	0·0001944 gr.
Bromine	0·0000515 ,,
Sulphide of sodium	0·067 ,,
Sulphate of soda	0·025 ,,
,, of lime	0·002 ,,
,, of magnesia	0·018 ,,
Chloride of sodium	0·018 ,,
,, of magnesium	0·014 ,,
Carbonate of lime	0·186 ,,
,, of magnesia	0·012 ,,
,, of soda	0·040 ,,
Silex	0·006 ,,
Carbonate of iron	0·013 ,,
,, of manganese	0·001 ,,
Sulphate of iron	0·007 ,,
Glairin	intermediate quantity.
Loss	0·017 ,,
Total	0·426 ,,

There is a regular omnibus service between Aix and Marlioz, or the distance can be covered on foot in less than twenty minutes. The altitude is 850 feet. Most comfortable apartments are found in the château and villa adjacent to the Etablissement, and the large picturesque park is highly attractive.

WATERS OF CHALLES.

Dr. Domenget, and subsequently our colleagues, Dr. Casalis and Dr. Roger, have by their detailed observations rendered Challes one of the most interesting spas in Europe.

These waters, highly impregnated with iodine, bromine, and sulphate of sodium, and situated at ten miles from Aix, have contributed greatly since their discovery to the general effect of our treatment. In my brief notice I can but indicate the composition of these marvellous springs which for the last thirty years have attracted the attention of *savants* and practitioners.

Challes waters rank first among sulphurous waters; large quantities are transported daily to Aix without any deterioration of the mineral or sulphurous element, and form an invaluable adjunct in our medication, as alteratives, depuratives, or solvents. Although somewhat unpleasant in flavour, patients, even children, rapidly surmount their original aversion to them. These waters increase when necessary the sulphurisation of Aix baths; and are likewise employed in injections and local applications.

Some years since, I presented to the Medical Society of Savoy the following note:—

"I recommend lotions of Challes water to patients of a cachectic type, who are losing their hair from pellagra, scrofula or herpes, and always with good results where the roots of the hair are intact."

Since that date further experience has confirmed my opinion. Subjoined is the analysis made by the distinguished chemist, Carrigou:

In 1000 grains of water:—

Sulphur (corresponding to free sulphuretted hydrogen)	0·0140 gr.
,, (corresponding to combined sulphuretted hydrogen)—	
,, as bisulphuret	0·0465 ,,
,, as monosulphuret	0·1128 ,,
,, probably as polysulphuret	0·0189 ,,
,, ,, as hyposulphite	0·0050 ,,
Sulphuric acid	0·0390 ,,
Silicic ,,	0·0092 ,,
Carbonic acid, free or as a bicarbonate	0·01162,,
,, ,, combined	
Phosphoric acid	0·00057,.
Nitric acid	0·0011 ,,
Chlorine	0·0870 ,,
Bromine	0·0016 ,,
Iodine	0·0089 ,,
Soda	0·4749 ,,
Potash	0·0057 ,,
Ammonia	0·0022 ,,
Lime	0·0856 ,,
Magnesia	0·0021 ,,
Alumina	0·00022,,
Iron	0·00039,,
Manganese	distinct traces.
Cobalt	slight ,,
Copper	abundant ,,
Lead	id.
Antimony	traces.
Arsenic	0·000007 gr.
Dialyzed organic matter	abundant.
Non-dialyzed ,,	abundant enough.
Total	1·023887 gr.

The temperature is 53° Fahr. Challes water is twenty times richer in sulphide of sodium than any of the other springs most favoured in this particular.

ST. SIMON WATERS.

This spa, neglected for many years, has recently resumed an important position in the therapeutics of Aix.

This water, unctuous to the touch, without smell, and of an agreeable flavour, has a natural temperature of 67° Fahr. The spring yields about 200,000 litres daily; and contains, according to M. Kramer's analysis, bicarbonate of lime, magnesia, potash, iron, sulphate of magnesia, alum, organic matters, &c.

When this water has been prescribed for irritability of the mucous membrane of the stomach, the sedative effects have always proved satisfactory.

Dr. Petrequin of Lyons has remarked its exciting influence on saliva, and I often observe its special effect on urine, and in cases of cystitis, when hot water douches over-stimulate the bladder, the waters of St. Simon rapidly calm the pain and inflammation. They are also efficacious in cases of rheumatism with visceral metastasis when there is a liability to gastric disturbances; and are very digestive as a dinner beverage. Four to six glasses daily assist the digestion.

WHEY CURE.

The combination of the whey cure with sulphurous treatment originated in Germany, and is chiefly met with in Alpine and Carpathian stations; more than twenty-five being exclusively adapted to consumption and chronic affections of the respiratory organs; in the Pyrenees also the happiest effects are produced.

In the parks of Aix and Marlioz, dairies exist where the patient drinks cows' or goats' whey according to the prescription, whilst taking exercise after baths and douches.

Acting as a mild aperient, whey succeeds in cases of catarrh of the stomach, not resulting from dyspepsia, and is invaluable in the digestive atony and constipation frequently met with at the commencement of the course.

As an alterative, it succeeds in herpetic affections; also in nervous erethism and irritability, produced by the early stage of the treatment.

As a means of nutrition, whey is a great resource to those enfeebled constitutions which, although incapable of digesting the fatty and caseine elements of milk, are benefited by its remaining component parts.

ELECTRICITY.

After witnessing the excellent results obtained by Drs. Onimus in Paris, and De Watteville in London, I ventured to employ at Aix the electric current; it is a useful adjunct to the course, owing to its stimulating and modifying effects on nervous, muscular and articular diseases; and is also a diagnostic agent in paralytic affections of doubtful nature. At first I encountered adverse criticism; but the value of continuous currents, which provoke neither spasms, organic disturbances, nor depression, is now recognised.

Continuous currents are successfully employed in cases of chronic rheumatism, muscular atrophy, arthritis, stiffness of the joints, and different forms of paralysis.

I was one of the first to employ continuous currents in cases of uterine tumour, and confidently assert that it has proved most successful in conjunction with general massage, the effect being to regulate the circulation which otherwise becomes impeded in the region of the tumour.*

* Cases of Dr. Brachet, published by Dr. Onimus, 1875.

PART II.

GENERAL INDICATIONS.

ROUTE BETWEEN PARIS AND AIX-LES-BAINS.

Fares :—First Class, fr. 71 c. 65 ; Second Class, fr. 53 c. 70.

The *Rapid Train* leaves Paris at 9 P.M. and arrives at Aix at 6.15 A.M., taking only 9 hours and a half.

Another rapid train leaves Paris at 8.55 A.M., arriving at Aix at 8.26 P.M. ; both convenient, without change of carriage.

There are many other trains in the day.

GENERAL COUP D'ŒIL OF THE COUNTRY.

How can I better describe the varied beauties of Aix than by quoting a portion of "A Pastoral in Green?" *

"The green mountains gird in Aix-les-Bains—the Alps of Grenoble and of Italy, possessing not only sympathetic outlines, but melodious colouring: sym-

* *All the Year Round*, August 31, 1878.

phonies in grey rock; chorales in green groves and bosky dingle, in gurgling stream, and in rushing river; dirges in dusky ravine and sombre forest, where nightingales chant. The purple tints, too, so charming among hills, cling to these Savoyard Alps willingly; purple shadows that gather as the afternoon sun lowers, the green tints showing through like shot-silk.

"The Lake of Bourget, shut out from the friendly little town of Aix by a long-shaped, envious, bowery hill, has much to do with the purple tints. Purple mountain-tints show up readily from sea, lake, or big broad river; a friendly understanding, as it were, between sky and water, reflected on mountain-sides. Those purple shadows are very dear to me, with that underlying richness of untrodden emerald lawn and rough woodland.

"Round the Lake of Bourget, and clinging about Aix, are belts and groves of sycamore and fir, walnut and ash, limes fragrant with blossoms, and lofty poplars shaking in the breeze; each mountain-tree forming a brick as it were in green walls, out of which the houses, and the little chalets, the Etablissement des Bains, the cathedral, the hotels, grow up like mushrooms in a water-meadow."

It is not in my province to eulogise the attractions of my country. Let me rather transcribe some of the graceful words of a well-known and most poetic writer :*—

* Right Hon. Lord Lamington, "Aix-les-Bains and Annecy"; *Nineteenth Century*, August 1883.

"For those who wish to combine the pursuit of health with a pleasant life and beautiful scenery, there are few places that offer so many advantages as Aix-les-Bains. It is only nine hours' journey from Paris. Those visitors who arrive in the early part of the season, before every favoured spot is thronged with bathers and tourists, will be able to combine all the good to be derived from the healing waters and soft soothing air with a life full of varied interest. Lamartine resided some time at Aix: it was here he wrote 'Raphael,' which may serve as an excellent guide-book for the district. He makes Raphael select Aix as a residence, because it combines the charm of the beautiful valley and fertile plain with the majesty of Alpine scenery. The district between Chambéry and Annecy does not exceed sixty miles, but these sixty miles are full of objects of interest; and the two lakes of Bourget and Annecy are not inferior in beauty to Maggiore and Como. How deeply Lamartine was impressed with this scenery may be perceived in every page of 'Raphael.' If Lamartine had identified himself with the lake of Bourget, Rousseau, in his 'Confessions,' has achieved the same result at Chambéry; while his early and later life passed at Annecy has added to the charm of that picturesque town; while his description of the lake invited so much attention to its beauty that many followed his example and built residences on its shores. Like Byron, Scott, and Burns, Lamartine and Rousseau have added to the charm of scenery which is exceptionally lovely. In 'Raphael,' Lamartine observes that

Nature, however grand and absorbing in interest, gains by its association with genius. 'How much,' he adds, 'does not Vaucluse owe to Petrarch; Sorrento to Tasso; Venice to Byron; Annecy and Chambéry to Rousseau and Madame de Warens!' And, we may now add, Aix-les-Bains to Lamartine!"

Unfortunately, the limits of my little work do not admit of quotations from many other authors who have helped to immortalise the beauties and charms of our beloved valley; but I must particularise the graphic and graceful pictures of H.R.H. Princess Beatrice, which appeared in the January number of *Good Words* of 1884, and the article by the Duchess of Rutland, which appeared in *The Queen* newspaper, January 1891.

THE TOWN OF AIX-LES-BAINS

Is situated at an altitude of 823 feet above the level of the sea, and 90 feet above Lake Bourget. Containing nearly 5800 inhabitants, it is built on the decline of the Revard chain of mountains about one mile from the lake. This cluster of hotels and houses, rising one above the other, presents a picturesque aspect from all points of view. The wide, well-planned streets abut on all sides on grand, shady avenues, affording delightful shelter during the sultry season. Almost all the hotels and many of the private houses are surrounded by gardens, which contribute greatly to the general salubrity. The Park, purchased with the Castle from the Marquis d'Aix

PLAN DE LA VILLE D'AIX-LES-BAINS

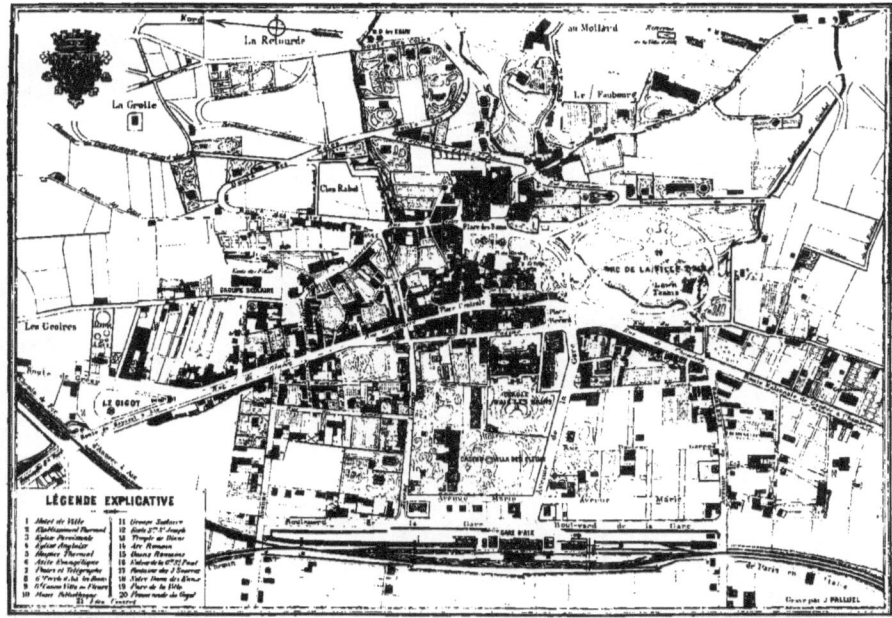

This page has been intentionally left blank

in 1869, was laid out as a public walk. The Castle now serves as the Post and Telegraph Office and Town Hall.

HOTELS—BOARDING-HOUSES— LODGING-HOUSES.

THE visitor finds superior accommodation at hotels, boarding-houses, and furnished lodgings, at fixed rates, varying from 9 to 18 fr. per day for board and lodging.

Considering that wine is included, this price is reasonable. Most adapted for perfect quiet are the furnished houses, about one hundred in number—generally with kitchens at the disposal of the lodgers, who have also the facility of taking their meals at any of the forty hotels or eighteen boarding-houses at fixed charges per day or meal. Many hotels and villas have been recently constructed on the heights overlooking the town and lake, offering a choice to those who prefer repose with glorious scenery to the advantages of the immediate proximity of the Baths and Casinos, with their numerous attractions. Large families requiring special accommodation are advised to write beforehand, as Aix is crowded during the season, particularly from June till the end of September.

CARRIAGES.

EXCELLENT landaus, victorias, and pony carriages can be hired by the month, day, or hour, or at special rates for excursions.

Carriages, with one or two horses, having tariff affixed, are always found at the railway station and in the streets. Omnibuses from the hotels meet all the trains at the terminus. To avoid difficulties with the coachmen, it is essential to fix beforehand the fare for excursions.

Chars-à-banc and breaks start at stated hours for the most attractive spots in the neighbourhood, at a very moderate tariff.

Donkeys and riding horses are always available.

STEAMERS AND BOATS.

STEAMERS leave the Grand Port daily at one o'clock, and conduct passengers round the lake, stopping at all the points of interest. Three times a week very large and comfortable steamers proceed from the same point to Lyons, returning the next day. A steamer can be secured for private parties by giving a few hours' previous notice. Sailing and rowing boats are hired by the day or hour for excursions or fishing.

TARIFFS.

OMNIBUSES.

	fr. c.
To Marlioz, and return	0 60
To St. Simon, single ticket	0 40
,, and return	0 60
To Gresy Waterfall, single	0 60
,, and return	0 80

TARIFFS.

CARRIAGES.

	fr. c.
Course in the town, one or two persons	1 0
,, ,, three or four persons	2 0
Drive out of the town (by the hour, without reference to the number of persons)—	
Carriage with one horse	3 0
,, ,, two horses	4 0

TARIFF ACCORDING TO DISTANCES.

From Aix to St. Innocent, Brison, and return—
Carriage with one horse	9 0
,, ,, two horses	12 0

Drive by the borders of the lake past Tresserves, Bon Port, and Terrenue, or Bon Port, Viviers, and Marlioz—
Carriage with one horse	8 0
,, ,, two horses	11 0

Drive to and from Bourget—
A carriage with one horse	11 0
,, ,, two horses	1 0

To and from the Moulin de Primes—
Carriage with one horse	9 0
,, ,, two horses	12 0

BOATS.

Two boatmen accompany boats containing six passengers—
To Haute Combe	9 0
Bourdeau	5 0
Bourget	8 0
Bon Port	4 0
Chatillon	14 0

DONKEYS.

By hour (each hour)	1 0
Half-day	4 0
Day	7 0

DIVINE SERVICE.

The Roman Catholic Church, rendered interesting by an ancient chapel within its precincts, is quite insufficient in size, and will shortly be replaced by one worthy in dimensions of the numerous visitors. Partly destroyed by fire in the couflagration of Aix in the fifteenth century, the edifice presents a variety of architectural epochs, from pure Gothic to the worst period of the Renaissance.

English Protestant service is regularly performed twice on Sundays in a charming little church recently enlarged to meet the requiremeuts of the yearly increasing numbers of visitors.

Presbyterian and French Protestant services take place also twice on Sundays in the large church annexed to the Anglo-French Hospital.

HOSPITALS.

To King Victor Amadeus of Savoy was due the inauguration of gratuitous baths with cheap board and lodging for indigent patients. Subsequently the Empress Josephine built and endowed a hospital in memory of the Baroness de Broc, who met with a tragic end in the cataract of Gresy. This hospital, at first insufficient in size, has been generously subsidised by native and foreign gratuities, among which the largest were contributed by Mr. Haldimand, Mrs. Boyd, and my father, many years Mayor of Aix, &c. &c.

Arc de Campanus a Aix

One hundred and twenty beds exist, and are generally occupied in summer.

An Anglo-French Protestant Asylum, established fifteen years ago, is supported entirely by voluntary contributions. The late M. Fournier, the French Protestant minister, and Mrs. MacViccar, an Irish lady, have collected large subscriptions in Great Britain and on the Continent, and visitors occasionally leave substantial additions to the funds. Patients treated in both hospitals are entitled to the gratuitous service of the Baths.

When patients can lay no claim to private charity they are admitted to both hospitals at a daily charge of two francs, on producing the necessary certificates of poverty signed by the Mayors of their communes.

THE ANTIQUITIES OF AIX.

The Arch of Campanus, almost entirely preserved appears to have been the entrance to the Thermes, though long erroneously regarded as a sepulchral arch of which purpose the inscriptions offer no sort of indication. The principal inscription—" L. Pompeius Campanus vius fecit "—may be supposed to designate the name of the author, and the other inscriptions the members of his family or his colleagues; or possibly the names of the administrators of the Thermes.

This arch is constructed, as well as the Temple, of huge blocks of stone supported without cement by their own weight. The western aspect presents eight niches, once probably occupied by the effigies of the persons whose names are inscribed at the base.

The Roman Bath, of octagon form, is one of the last specimens of Roman architecture. The bricks employed in its construction bear the mark of Clarianus, a manufacturer residing in Vienne (Dauphiné), and are similar to those found in the Roman ruins at Lyons, thus proving that the Roman constructions at Aix and Lyons date from the same epoch—either the reign of Augustus or of Tiberius. This Roman bath is situated in the cellars of the Pension Chabert, and in the same house is a Roman sun-dial originally found in the Roman bath. Fragments of ancient marble statues were discovered in the excavations around the bath, as also amphoræ, and many coins dating from the two first Christian centuries.

The Temple of Diana.—This edifice, hedged in by modern buildings, denotes by its elegant simplicity the taste of Grecian architecture improved by Roman progress. Unfortunately, one façade opposite the Baths is destroyed; the other façades are composed of huge blocks of stone, piled up over each other. Archæologists attribute this construction to the reign of Tiberius, some believing that it was dedicated to the worship of Venus, others to that of Diana, according to local traditions.

Museum.—A staircase of the sixteenth century of the purest Renaissance style leads to the interior of the Temple of Diana, with a Museum on the first-floor, containing many objects of local interest and great antiquity, the gift of Count L.

Casino d'Aix vu des Jardins

Lepic. The price of admission is one franc. In the same edifice is a Library, where all the works and maps referring to the history of the country are collected.

AMUSEMENTS.

We have two large casinos—the **Casino Ancien Cercle,** a club for ladies and gentlemen personally introduced, was built by subscriptions forty years ago. The shares, entirely taken up in Savoy, receive no interest, the profits being applied to the improvement of the club. Thus, five years since an elegant and commodious theatre, and a spacious hall decorated with Salviati's mosaics, were constructed.

Twice a day, in addition to other amusements, a band plays in the gardens. In the evening there are alternate performances of opera and concerts executed by Colonne's celebrated orchestra from the Chatelet in Paris; once a week fireworks and a ball, and occasionally special performances of plays by the first Parisian artists. The Reading-room is provided with the leading newspapers of all countries, and an excellent restaurant is attached.

CASINO.

	fr.	c.
For the season—		
One person	40	0
Husband and wife	65	0
Father or mother with one child, under twelve years old	65	0
Father and mother, with one unmarried child	85	0
Family, composed of father, mother, and several unmarried children	100	0
Children under twelve years old	5	0
Single entry by day	3	0

The Villa des Fleurs offers the same amusements, and is also much frequented, the spacious gardens proving very attractive. The directors of both these establishments are most courteous and attentive to foreigners.

VILLA DES FLEURS.

	fr.	c.
For the season—		
One person	40	0
Two persons	70	0
Each extra person	20	0
Subscription for a fortnight at the rate of two francs per day.		
Single entry by day	3	0

Pigeon Shooting.—Every year shooting matches, in which the chief champions of Europe take part, occur at Marlioz, about one mile from Aix.

Races.—A local company has undertaken the establishment of annual races at Marlioz, a picturesque adjacent spot.

WALKS.

Roche du Roi—an ancient stone quarry, commanding the finest view of the town and lake. Ascending by the Splendid Hotel, and returning by the avenue of Marlioz, this walk occupies about half an hour; morning is the best time for witnessing the effects of light and shade.

Boulevard des Cotes.—Turning to the left of the Etablissement, follow this new boulevard by a gentle ascent, with an uninterrupted view of the town, and of the furthest extremities of the lake and mountains. On both sides of the road numerous villas are built, and others are projected. If not too fatiguing, it is advisable to proceed as far as Notre Dame des Eaux, a most commanding situation; otherwise, a few minutes' walk leads direct to the Park and town.

Park of Marlioz.—Those who are not strong enough to ascend the heights, enjoy a walk of about twenty minutes' duration to Marlioz, through one of those splendid avenues of trees for which Aix is justly celebrated.

St. Simon, Pont Pierre, Chateau de Syllan.—Passing the cemetery by the Geneva road, a short walk leads to St. Simon, renowned for its mineral alkaline spring, and for a large plantation of roses, offering a varied choice to the intending purchaser. If not tired, proceed till you reach a stone bridge (Pont Pierre), turn to the left, and a gentle ascent through a vineyard abuts on the wood of Touvieres and the Castle of Syllan. You will be amply rewarded by the grand panorama of the Alpes Dauphinoises, the mountain of the Grande Chartreuse, the general view of the town and lake, and, above all, of the plain, the seat of a great battle between the Romans and the Allobroges, where 120,000 men fell victims.* Return by St. Innocent, the abode of the Angora rabbits, and the Grand Port. This walk, occupying about two hours, may be performed on foot or on donkeys.

Maison du Diable.—Close to the road of the Petit Port is the north end of the Hill of Tresserves,

* Cabias, 1668.

on which stands the Maison du Diable—so named after a popular legend—the property of the Hon. Lady W——, who for many years exercised with her late husband profuse hospitality towards the inhabitants and visitors of Aix.

Bois Lamartine, situated on the decline of the Hill of Tresserves, was illustrated by the most sympathetic of our poets, to whom we owe the glorious poem of the "Lake."

The Grand Port.—A splendid avenue to the left of the Gigot leads in about twenty minutes to the Grand Port, the starting-point of the steamers, with a general view of the Lake and its surroundings.

The Petit Port.—Still nearer to Aix is the Petit Port, whence the lake is viewed from another aspect. The distance on foot is about a quarter of an hour from Aix. A new canal is projected to bring the waters of the Lake into the centre of the town, thus enabling visitors to start in boats direct from the hotel.

The Hill of Tresserves, at the south of Aix, commands on one side the Bois Lamartine and the town, with its mountain background; and on the other, through richly wooded slopes, an enchanting view of the lake and surrounding hills. Some of the villas are occupied by English residents. Returning by the borders of the lake, an hour and a half is agreeably occupied. If Lamartine's description has rendered Tresserves famous, the acquisition of lands by the Queen of England in 1886 will make it illustrious in all ages.

Mouxy, Trevignin, Clarafond. — Leaving the town, and passing behind the Etablissement, the visitor arrives by an easy ascent at the village of Mouxy, thence to the left at the village of Trevignin, and to the right at that of Clarafond. All these roads are beautifully wooded, and the atmosphere becomes more bracing as we ascend.

DRIVES.

Waterfall of Gresy.—The most frequented excursion by carriage or train is to the Cascade of Gresy, about twenty minutes' drive from Aix, where the meeting of two rivers produces a glorious effect. At the distance of two minutes' walk from the high road, a little steam tender is in waiting to take the visitor through a narrow and highly picturesque gorge, a mile in length, presenting no danger.

Tower of Gresy.—I advise the visitor to ascend to this Tower, an ancient building, dating from the end of the eleventh century, whence the view is magnificent. According to some historians, this was the site of the Cemetery during the Roman occupation of Aix, and of the Christian Cemetery in the sixth century. The Donjon Tower was built in the twelfth century. Great interest is derived from a careful study of the ancient inscriptions at its base.

Moulin de Prime.—Continuing this drive, a shady and charming road, most enjoyable during the hot season, leads to the Moulin de Prime, about five miles distant from Aix.

Grotte de Bange.—Farther on is the Grotte de Bange, well worth a visit—a torrent (Le Chérant) following the entire length of the road refreshes the atmosphere; gold-dust is found in the bed of the river.

The so-called Grotto, an immense cavern, about fifteen minutes from the high road, with a lake at its extremity, is difficult of access.

The drives occupies from four to five hours.

Tour de Lac.—To obtain a general idea of the country, the best plan is to follow the right border of the lake, and return by Tresserves or Viviers.

The drive occupies an hour and a half.

Chateau de la Serraz.—Turning to the right of Terrenue, at the head of the lake, a delightful drive through groves of pines and chestnuts leads to La Serraz Castle, whence the scenery is lovely.

Chateau de la Motte.—The straight road from Terrenue conducts to the Chateau de la Motte. The splendid park resembles in many respects the far-famed English parks, seldom met with in our country.

These two expeditions occupy three or four hours.

Le Bourget et le Col du Chat.—The village of Bourget, about fifty minutes from Aix, is situated on the opposite side of the lake.

The well-preserved ruins of an old castle built in the thirteenth century by the Counts of Savoy, the village church contiguous to the dilapidated remains of another ancient castle, are the chief objects of interest. A little farther on, a winding carriage-road leads by an easy ascent, in less than an hour, to the Col du

Chat. By this road Hannibal achieved his renowned passage across the Alps, two hundred and fifteen years before the Christian era.

This excursion cannot fail to charm and interest the visitor.

Castle of Bourdeau.—The straight road from Bourget conducts to the Castle of Bourdeau—a hunting-box of the old Counts of Savoy constructed in the eleventh century. Frequent repairs at different epochs give to this castle an air of originality singularly picturesque when viewed from the lake.

The commanding aspect of the terrace, at an altitude of 900 feet, deserves special comment. The grand saloon contains some old Italian pictures of a certain merit.

Haute Combe, the ancient burial-place of the House of Savoy, and almost the only spot there still belonging to that Royal race, dates from the twelfth century, when Amadeus the Third bestowed the site on the Abbot of Clairvaux. The Abbey was doomed to destruction in 1792 and transformed into a pottery manufactory; but was completely restored and redecorated in 1824 by King Charles Felix, and is now occupied by some Bernardine monks.

In the church are more than three hundred statues, many old pictures, mausoleums, and simple tombstones. The momuments of King Charles Felix, buried in 1831, and of his Queen, Mary Christine, buried in 1849, are the most interesting. No admission is granted to the cloister. The private apartments of the Kings of Savoy may be visited.

In the hottest weather there is a refreshing breeze on the lake. Invalids and all who walk with difficulty enjoy the entrancing views around without leaving the boat, which, as a change, is recommended to all as a delightful mode of transit.

Castle of Chatillon.—A gay and smiling road—a miniature Corniche, winding at the base of the mountain by the borders of the lake, leads in an hour and a half from St. Innocent to Chatillon, enabling the visitor to survey the stupendous railway works constructed by an English Company thirty years ago. The castle, standing on a platform at the summit of a hill, commands the view of the entire lake, the terrace and the ancient tower, whence not only the lake, but the Rhone Valley are plainly distinguished. In this castle, Pope Celestin IV. was born, at the end of the eighth century. In hot weather, it is best to proceed by train to Chindrieux station; the little Corniche road is exposed to the sun, and affords no shelter.

Drive to the Rhone by the Canal of Savière.—This bright and attractive route, highly recommended to people too nervous for boat excursions, follows the canal through which the waters of the lake flow into the Rhone by a bridge recently constructed over the canal of Savières. Carriages arrive direct by a steep but practicable route at Haute Combe—a great advantage for timorous people not venturing in boats.

Val de Fier.—This road, about nine miles in length, was constructed twenty years ago, through mountain gorges hitherto deemed impassable, and

Château de Châtillon.

follows the course of a deep river (the Fier), through which the waters of Lake Annecy flow into the Rhone.

There is ample time for this excursion between luncheon and dinner. Take the train from Aix to Rumilly (an ancient and interesting town), drive to Seyssel by the Val de Fier, and return by train from Seyssel to Aix.

Mont Joìgny (1548 metres).—This is the most interesting of excursions, and may be undertaken even by delicate persons, who drive from Chambéry to the tunnel of the Pas de la Fosse in two hours. The views are splendid, both in the valley which extends from Chambéry to Annecy and in the valley of Isère. Arrangements must be made to drive as far as the Cantine, whence good walkers ascend a path for about a quarter of an hour, and are rewarded by a view of the glorious panorama of Mount Blanc. Three-quarters of an hour's further walk leads to the summit of Mount Joìgny, whence the views are incomparable. To accomplish this trip easily, it is advisable to start by train from Aix at about 10 A.M., to breakfast at the Hotel de France at Chambéry, whence carriages are engaged, and to return to dine at Chambéry or Aix. This delightful excursion is especially recommended to all who desire to form a true idea of the beauties of Savoy.

Valley of the Beauges.—A continuation of the road past Gresy, the Moulin de Prime, and the Grotto of Banges, leads to a fertile Alpine valley, rich in vegetation, and abounding in glorious points of view. The excursion occupies an entire day, and, by way of economy, a large party should be arranged to share the expenses. Although food may be obtained at the inns, it is more prudent to take a hamper of provisions. Starting in carriages or brakes at seven in the morning, the best halting-place for breakfast is the Châtelard, a charming little village at the base of a steep rock. Drive on through a picturesque road, penetrating the very heart of the Alps to St. Pierre de Albigny, whence the train occupies an hour and a half in returning to Aix; or Aix may be reached through the Deserts, or by the Chartreuse des Aillons; each road offering so many attractions, that it is impossible to indicate the best mode of surveying the glorious Alpine effects.

Castle of Miolan.—If the return by way of St. Pierre d'Albigny is selected, the Castle of Miolan, an ancient State prison, is well worth visiting.

When going direct by railway from Aix, carriages must be previously ordered from the hotel-keeper of St. Pierre d'Albigny.

Pierre Chatel.—This excursion must not be missed; either carriages or boats are available.

The steamer starts from the Grand Port, crosses the lake and the canal of Savières, descends the Rhone, which is bordered on both sides by picturesque hills and mountains, and in two hours arrives at Pierre Châtel — at one time one of the most formidable forts between France and Savoy, but now occupied only by a small garrison, to whom the beauties are less evident after long months of incarceration than to the passing tourist. In the caves may be seen the vestiges of the convent, founded in 1362, by Count Fert of Savoy, the most popular and perfect chevalier of the Middle Ages. The Order of the Annonciade occupies so important a place in the history of the House of Savoy, that a brief account of its foundation will certainly be welcome. It was established by Count Fert on returning from his victorious campaign. The number of dignitaries was not to exceed twelve, and they were to be buried in the convent of Pierre Châtel, where an equal number of monks were to pray for the success of the members of the Order on the field of battle, and for the repose of their souls after death. The distinguishing decoration of the Order was a dog's collar, whence depended a medallion with the device of Fert, interlaced with two true-lover's knots.

The Grande Chartreuse. — This excursion is frequently made in one day, but when two days can be spared proves far more agreeable. By starting after the bath or douche at ten or eleven in the morning, the return to Aix in time for the bath the next afternoon is easily effected, as the Etablissement remains open until five. It is advisable to engage a landau for four persons. The gentlemen of the party sleep at the monastery, and may attend the morning service; whilst the ladies find every comfort at an adjacent inn, and may rest or witness the splendid effect of the rising sun. The combination of the ideal and the mystic with the beauties of Nature endows this excursion with special attractions, intensely appreciated by all visitors.

Lake of Aiguebelette — Spas of La Bauche.—At the south extremity of the Mont du Chat, behind the mountain of L'Epine, a little lake, nearly three miles long and a mile and a half broad, glistens like an opal in the midst of pine groves. A train from Chambéry deposits the tourist at the station of Aiguebelette, close to the lake, whence it is easy to visit the iron spas of La Bauche. Half a day suffices for the whole expedition.

MOUNTAIN ASCENTS.

Dent du Chat.—The summit of the Dent du Chat is the most frequented mountain ascent in the environs of Aix. Delicate people and invalids must not attempt to climb to the most elevated peak, five thousand feet high. Carriages drive to the inn, whence—as all along the road—a magnificent view is obtained. In thirty minutes a steep accessible route enables the tourist to enjoy the full view over the Alps, Mont Blanc, and the Rhone Valley; the expedition occupies four or five hours.

To shorten the distance the lake is occasionally crossed in boats. From a medical point this mode cannot be recommended; fatigue and exhaustion predispose to chills, with all their dangerous results; whilst in a carriage sufficient wraps avert all risk.

The Revard.—Some months since the Revard Railway was commenced, and it will be open for traffic for the season of 1892; the length is nine kilometres. Starting from the Park it will reach the summit of Mount Revard at a height of 1545 metres above the sea level. The distance will occupy one hour. There will be five stations, the Park, Mouxy, Pugny, Prejapert and Revard. The works of art will include a viaduct with six arches, and a tunnel sixty metres long. The carriages will be of one class only, but the Company has the right of adding reserved places if advisable.

The Medical Corps of Aix originated the project of bestowing on our country this fresh attraction twelve years ago; M. Richenbach, a Swiss engineer, gave his professional assistance. Since then MM. Barbier and our friend Dr. Monnard have worked so assiduously on technical and medical considerations that a Company has been induced to undertake the construction of the Line. Very shortly we shall open to our patients this station in the immediate vicinity, in all ways adapted for an after-cure.

Aix Revard will possess a Sanatorium which in all points can vie with those of the Swiss Alps, the Engadine, Styria, the Black Forest, and those more distant of Norway, the Andes, the Rocky Mountains, and Himalayas.

The Dent du Nivolet.—The best way to the Nivolet is by train or road to Chambéry; thence two hours' drive to the village of the Deserts and a walk of two hours lead to the summit (5070 feet). The view of the Chambéry valley is unequalled.

Semnoz Alps, otherwise called Righi of

Savoy.—At four hours' drive from Aix is the foot of Mont Semnoz. The ascent occupies two hours on foot or donkey. By sleeping at the excellent hotel-châlet—at an altitude of 5600 feet—enjoying the entire panorama of the mountains and lakes of Savoy and Switzerland, the varied effects of sunrise and sunset may be witnessed. The projected railway will remove all existing obstacles, and render the ascent of the Semnoz practicable to the most delicate persons.

Tower of Cessens.—To reach this very ancient tower the Geneva road is followed as far as Albens, whence an ascent by a good carriage-drive, in aspect resembling Scotch scenery, leads to the summit of the mountain; an excellent view of the valley of the Rhone and the lake may be here enjoyed. Return by Albens, St. Girod, the Castle de Loches, and Gresy. This excursion is easily performed between luncheon and dinner.

CHAMBÉRY.

The former capital of the Dukedom of Savoy has preserved, amidst innumerable varieties of architecture denoting changes of dynasty, its ancient type. As seat of the Prefecture of the Department of Savoy, greater animation prevails than in ordinary towns occupied by twenty thousand inhabitants. The new streets, wide and clean, contain excellent shops; the Palace of Justice and the Town Hall are singularly fine edifices. The majority of the hospitals and asylums are due to the benevolence of the Comte de

Boigue, a native of the town, who returned from the East Indies with a colossal fortune; in recognition of his services, the municipality erected in his honour a monument of doubtful taste surrounded by effigies of elephants, which occupies a prominent position in the town. The old Castle, a most interesting monument, built in 1232, has been so often partially burnt and restored that it now presents every variety of architecture. Formerly occupied by the Governors of Savoy, it now serves as the Prefecture. From the summit of the tower a general *coup-d'œil* of the enchanting surroundings amply repays the fatigue of the ascent. In the Sainte Chapelle, the porch in Renaissance style, and some ancient frescoes and stained glass windows, are remarkable. The little Gothic chapel is also interesting. The Museum contains a complete collection of antiquities discovered in the lake; of old coins; and of the costumes formerly worn in different parts of Savoy.

Les Charmettes.—The remembrance of Jean Jacques Rousseau and Madame de Warens enhances the interest of this picturesque cottage, half an hour's drive from Chambéry, commanding some fine points of view.

Challes.—The Etablissement, the Casino, and the waters, more highly impregnated with sulphur than any others in the world, render Challes very interesting; the hotel of the old Castle is also curious; and on the terrace, where dinners are served, may be enjoyed an admirable view of the setting sun, as it sinks behind the range of the Alps.

The Bout du Monde.—This charming waterfall, issuing from the Doria Gorge at the foot of the rock of Mount Nivolet, works an important paper mill; and is little out of the direct road from Chambéry to Challes.

All the drives round Chambéry are distinguished by the richness and variety of the vegetation, and the magnificent points of view.

Gorges du Fier.—Leaving the Annecy train at Lovagny—an hour and a quarter from Aix—the Gorges du Fier are at a short distance; invalid chairs are available. The gallery, 260 metres in length, is protected from all risk by a balustrade a metre in height; a mild and soothing temperature pervades. The torrent, pursuing its headlong course amidst rocky crags beneath the gallery, presents an aspect of striking grandeur.

The Château of Montrotiers—close to the Gorges—dating from the fourteenth century, resembles, in miniature, the splendid Castle of Pierrefonds. Once a baronial residence, Montrotiers has been recently purchased by a gentleman from Lyons.

ANNECY.

The railway transit from Aix to Annecy occupies one hour and twenty minutes. Population, 12,000 inhabitants. Situated at the foot of Mount Semnoz, and at the extremity of the lake bearing its name, Annecy possesses two excellent hotels, and carriages and boats of all descriptions are available.

Château d'Annecy vu du Lac

For further details I again gladly refer to Lord Lamington's vivid and graceful description:*—

" After a long experience of travel, seldom have I seen any place combine so much to charm and interest as Annecy. The lake not only flows up to the town, but through canals which are crossed by narrow bridges; and on either side are quaint old houses, such as Prout loved to sketch: which, although many of them of wood, have remained unchanged for centuries. Those who love symmetry must not penetrate into the streets and lanes of Annecy; but the artist may find subjects for his brush occupy many a day: every turn presents a fresh object of delight to the student in pencil or pen. It is strangely diversified in appearance. There are stately houses in the old Venetian style, with balconies of highly finished ironwork, and decorated architraves, where old families still reside in dignified retirement. Commanding the town is the quaint old castle of the Dukes of Genevois-Nemours, dating from the fourteenth century—with its towers and massive keep, its ramparts and battlements, which so frequently and successfully defied the power of France, and averted that conquest which now has been achieved by purchase. Annecy is a Bishopric and Prefecture; and as troops are always garrisoned there, there is enough animation to add to its mediæval interest. The dignity of a Prefecture has led to the formation of delightful gardens on the

* " Aix-les-Bains and Annecy": Lord Lamington, in *Nineteenth Century*, August 1883.

shores of the lake, which is about twelve miles long by three wide, and is surrounded on three sides by Alpine snow-covered mountains. Thus Annecy adds to the loveliness of Como the grandeur of Lake Leman.

"Many illustrious men have found their rest in retirement on its shores. Eugene Sue, Custine, and Rousseau passed the latter years of their lives there. 'The Maisonnette de Chavoires,' now called the 'Maison de Rousseau,' where he dwelt so many years, still exists, although falling into ruins.

"The Académie Florimontaine held its sittings in the ancient episcopal palace, where François de Sales lived. No name was so honoured and loved as this Apostle of the Alps. He died in 1622, and was buried in the beautiful cathedral. In such estimation was his memory held by the people, that when the city was taken by the French in 1630, one of the six articles of capitulation was that the body of François de Sales should never be moved from the city.

"Even those who take little interest in spots associated with genius such as Rousseau's, or in noble lives such as Saint François de Sales' (he was canonized in 1658), or who only visit Annecy for change of air or love of beautiful scenery, will be grateful to be invited there. There are few spots where the love of retirement can be more pleasantly indulged, and there is sufficient movement to prevent the painful sense of solitude. It is very charming to sit in the beautiful park and watch the lights and shadows on

Intérieur du vieux Annecy.

the rich wooded hill-sides, while far beyond are seen the mountains of the Val d'Isère; and still further distant the snowy summits of the Dauphiné Alps. Another advantage is that Annecy, although less than two hours distant from Aix-les-Bains, is comparatively little known to tourists. It is fervently to be hoped that no sulphurous spring may be discovered, and that it may remain a little city to flee to from the *balnea strepitumque* of Aix. After a few days spent there the invalid will return with increased energy to complete his treatment."

www.ingramcontent.com/pod-product-compliance
Lightning Source LLC
Chambersburg PA
CBHW030343170426
43202CB00010B/1218